Composite Media with Weak Spatial Dispersion

Composite Media with Weak Spatial Dispersion

Constantin Simovski

PAN STANFORD PUBLISHING

Published by

Pan Stanford Publishing Pte. Ltd.
Penthouse Level, Suntec Tower 3
8 Temasek Boulevard
Singapore 038988

Email: editorial@panstanford.com
Web: www.panstanford.com

British Library Cataloguing-in-Publication Data
A catalogue record for this book is available from the British Library.

Composite Media with Weak Spatial Dispersion

Copyright © 2018 Pan Stanford Publishing Pte. Ltd.

All rights reserved. This book, or parts thereof, may not be reproduced in any form or by any means, electronic or mechanical, including photocopying, recording or any information storage and retrieval system now known or to be invented, without written permission from the publisher.

For photocopying of material in this volume, please pay a copying fee through the Copyright Clearance Center, Inc., 222 Rosewood Drive, Danvers, MA 01923, USA. In this case permission to photocopy is not required from the publisher.

ISBN 978-981-4774-83-3 (Hardcover)
ISBN 978-1-351-16624-9 (eBook)

Contents

Preface	xi
1 General Introduction to Weak Spatial Dispersion	**1**
1.1 Natural and Composite Media with Weak Spatial Dispersion	1
1.1.1 Chirality	2
1.1.2 Artificial Magnetism	6
1.1.3 About Non-Local Description of Media with Chirality and Artificial Magnetism	8
1.2 Homogenization	11
1.2.1 Homogenization in General	11
1.2.2 About Bulk Homogenization	14
1.2.3 About Homogenization of Interfaces	15
1.2.4 How to Apply the Homogenization Model	17
1.2.5 About Homogenization of Media with Weak Spatial Dispersion	18
1.3 Weak Spatial Dispersion versus Strong Spatial Dispersion	20
1.4 Continuity and Locality	27
1.4.1 Relations by Kramers and Kronig and Their Violation	27
1.4.2 Lorentzian Dispersion of Material Parameters	33
1.5 About this Book	35
1.5.1 How Our Theory Is Presented	35
1.5.2 Peculiarities of Notations in this Book	37
2 Quasi-Static Averaging of Microscopic Fields and the Concept of Bianisotropy	**41**
2.1 View on Weak Spatial Dispersion in the Available Literature	41

2.2	Introduction to Quasi-Static Homogenization	45
2.3	Classical Derivation of the Clausius–Mossotti–Lorenz–Lorentz Formulas	46
2.4	CMLL Formulas in Optical Theories	52
	2.4.1 Homogenization of Semi-Infinite Crystals	52
	2.4.2 On the Effect of Surface Polaritons	59
	2.4.3 On the Impact of Randomness	63
	2.4.4 A Bit More about the Anisotropy	67
2.5	Maxwell Garnett Model for Dielectric and Magneto-Dielectric Composites	68
	2.5.1 Maxwell Garnett and His Studies of Metal Glasses	68
	2.5.2 Maxwell Garnett Model for Magneto-Dielectric Composites	71
2.6	Bianisotropic Media	73
	2.6.1 Introduction to Bianisotropy	73
	2.6.2 Chiral Media, Omega-Media and Their Microwave Realizations	78
	2.6.3 Magnetoelectric Coupling in Metal Bianisotropic Particles	80
	2.6.4 Maxwell Garnett Model for a Medium with Both Chirality and Artificial Magnetism	84
2.7	Some Restrictions of Our Study Subject	86
	2.7.1 Why We Do Not Consider the Condensed Composites of Complex-Shape Metal Particles	86
	2.7.2 On Composites of Dielectric Particles	89

3 Multipolar Theory of Weak Spatial Dispersion — 93

3.1	Preliminary Speculations	93
3.2	Main and Auxiliary Vectors of the Macroscopic Electromagnetic Field	95
3.3	Multipole Expansion of the Macroscopic Polarization Current	99
	3.3.1 Microscopic and Macroscopic Multipole Densities	99
	3.3.2 Attempts to Avoid Multipoles in the Model of Weak Spatial Dispersion	102
	3.3.3 On the Advantages of Multipoles	105

	3.3.4	Multipole Moments	106
	3.3.5	Polarization Current in Media with Weak Spatial Dispersion	109
	3.3.6	Electric and Magnetic Polarization Currents	110
3.4	Material Equations for Media with Weak Spatial Dispersion		112
	3.4.1	Non-Covariant Constitutive Equations	112
	3.4.2	Why Non-Covariant Equations Cannot Be Used in Boundary Problems	117
	3.4.3	Material Equations Covariant in the First Order of WSD	122
	3.4.4	Material Equations Covariant in the Second Order of WSD	124
	3.4.5	Some Special Cases of MEs in Media with WSD	128
	3.4.6	Misinterpretation of Multipolar Composites Treated as Magneto-Dielectric Media	130
3.5	On Two Equivalent Approaches to WSD		134
3.6	Some Preliminary Conclusions		139
3.7	How the Non-Locality of the Particle Response Results in WSD		140
	3.7.1	A Qualitative Microscopic Analysis of an Explicit Example	140
	3.7.2	Spatial Dispersion Expressed by the b-Parameter	146
	3.7.3	A Relation between the Dipole and the Quadrupole Susceptibilities	148
	3.7.4	On the Hierarchy of Multipoles for Densely Packed Media	151
	3.7.5	Generalized Maxwell Garnett Model for Media with WSD	154
3.8	WSD in a Non-Resonant Array of Dielectric Spheres		160
	3.8.1	About Artificial Magnetism of Dielectric Spheres	160
	3.8.2	A Simple Cubic Lattice of Dielectric Spheres: Theory	162
	3.8.3	A Simple Cubic Lattice of Spheres: Calculations	165
3.9	Conclusions of This Chapter		169

4 Revision of the CMLL Formulas and Boundary Conditions for Thin Composite Layers — 173

- 4.1 Retardation Effects in the CMLL Formulas — 173
  - 4.1.1 Interaction of Non-Resonant Particles — 173
  - 4.1.2 Averaging of the Reference Particle Field — 178
    - 4.1.2.1 Static case — 178
    - 4.1.2.2 Dynamic case — 180
  - 4.1.3 Modified CMLL Formulas — 182
- 4.2 On Boundary Conditions beyond the Metamaterial Frequency Region — 186
  - 4.2.1 Preliminary Remarks — 186
  - 4.2.2 Drude Transition Layers as Effective Boundaries — 188
  - 4.2.3 Drude Transition Layer for Ultimately Thin Composite Slabs — 193

5 Homogenization of Metamaterials with Artificial Magnetism — 199

- 5.1 Dynamic Averaging — 199
  - 5.1.1 General Approach to Dynamic Averaging — 199
  - 5.1.2 New Definitions of the Mean Fields and Polarizations — 201
  - 5.1.3 Averaging of Maxwell's Equations — 205
  - 5.1.4 Generalization of the Model — 209
- 5.2 Effective Material Parameters of the Resonant p-m-Lattices — 213
  - 5.2.1 Wave Number of a p-m-Lattice — 213
  - 5.2.2 Wave Impedance of a p-m-Lattice — 220
  - 5.2.3 Effective Material Parameters of an Infinite Lattice — 221
    - 5.2.3.1 A two-phase lattice of dielectric spheres — 223
    - 5.2.3.2 A lattice of omega dimers — 229
  - 5.2.4 Standard Retrieval of Bulk EMPs from the Scattering Matrix of a MM Slab — 234
  - 5.2.5 On So-Called Bloch Lattices — 239
  - 5.2.6 A Correct Retrieval Procedure for Bloch's Metamaterial Lattices — 243
    - 5.2.6.1 Locality of retrieved parameters — 245

		5.2.7	Retrieval of the Transition Layer Parameters	248
	5.3	Recent Progress		253
		5.3.1	Effective Surface Sheets Instead of Transition Layers	253
		5.3.2	Bianisotropic Parameters of Non-Bianisotropic Materials	256
			5.3.2.1 Resonance band: optical losses versus Bragg resonance	256
			5.3.2.2 An effectively continuous resonant lattice without optical losses	260
			5.3.2.3 A lattice with low optical losses	265
6	**Conclusions**			**269**
References				273
Index				289

Preface

In this book, a modern theory of so-called *weak spatial dispersion* (WSD) in composite media of optically small inclusions without natural magnetism and optical nonlinearity is presented. WSD manifests in two important phenomena called *bianisotropy* and *artificial magnetism*, whose microscopic origin is thoroughly studied. The theory of this book is applicable to natural media with WSD such as chiral materials. However, emphasis is laid on artificial media allowing us to engineer needed electromagnetic properties. A homogenization model of effectively continuous media with general multipole response is presented. The model is especially important for so-called *metamaterials* in which artificial magnetism can be a resonant phenomenon and may result in the violation of Maxwell's boundary conditions and other challenges.

Partially, this book represents the English version of my Russian book [1] published in 2003. That book was one of the first monographs in which electromagnetic metamaterials were considered. At that time, this term was identified with composites targeted to implement the so-called *perfect lens* suggested in 2000 by Sir John Pendry in the work [2]. That perfect lens was a layer of a medium with permittivity ε and permeability μ both equal to (-1). Such media do not exist in nature and should have been engineered. If it were possible to implement exactly at least in a narrow frequency band, it would have offered us the reproduction of a scattering object with all its slightest subwavelength details in its far-field zone. This hypothetic regime was called the *perfect imaging*. In fact, a hypothetic medium with permittivity ε and permeability μ both equal to (-1) was theoretically introduced by Victor Veselago, who claimed it to be a prospective composite, in which several phenomena impossible in natural materials and predicted in [3] will be observed. In his pioneering work, Veselago has not revealed the perfect imaging effect. It was pointed out only in 2000—by Pendry.

For researchers, including me, in 2003 the word metamaterials (MMs) designated composite magneto-dielectric media whose effective permittivity and permeability, though complex, had negative real parts (not obviously equal to (-1)). Such composites, now called *doubly negative media*, have been designed since 2000 by several scientific groups. The first one was that of David Smith. The purpose of these works was to approach the target—material parameters $\varepsilon = \mu = -1$. The exact implementation of such material parameters is hardly possible (see [4]), but the hope to approach this idealized target closely enough was strong.

Media with *weak spatial dispersion* (WSD) described in [1] could be doubly negative. However, in 2003 the spatial dispersion in doubly negative media (namely, the impact of the magnetic resonance) was not sufficiently studied. Therefore, in [1], doubly negative metamaterials were discussed only briefly. Moreover, I could not concentrate on metamaterials because the main purpose of my monograph was educational. My friend Alexander Sochava taught a course *Electrodynamics of composite media* for master students and PhD students at Polytechnic University of St. Petersburg. Initially, in this course both artificial magnetism and bianisotropy phenomena were explained without homogenization models because the course followed to the famous book by L. D. Landau and E. M. Lifshitz [5]. In this book, the description of artificial magnetism and optical activity (the special case of bianisotropy) is presented without any homogenization model. The most curious students of Alexander were eager to understand how and why these two phenomena arise in a real medium consisting of molecules, but could not understand it from their study book. However, more relevant books in this field were absent and the lecturer could only ask such students to read a dozen of difficult scientific papers.

In my monograph [1], I tried to present the homogenization model for composites with WSD in a comprehensive way suitable for these master students. I have done my best in order to keep my explanations similarly comprehensive in the present English book. This book even in what concerns the non-resonant composites differs from [1] significantly. First, new papers on WSD have been published since 2003. Second, both the terminology and the scientific priorities have changed. Perfect imaging is not a hot topic anymore and MMs are now defined differently. MMs are, nowadays, *effectively continuous composite structures with unusual and useful electromagnetic properties* – properties not observable in nature and suitable for practical applications. This definition can be found, for example, in [6] and is probably the most popular definition of MMs. Although this definition is more general than the old one, still it does not cover all composite media with WSD. Therefore, the major part of this book refers to the non-resonant composites that cannot be referred to as MMs. However, the chapters explaining the artificial magnetism and bianisotropy in composites and in some molecular media are instructive for better understanding these phenomena.

The last chapter of this book is dedicated to the homogenization and electromagnetic characterization of MMs. Probably, it is the most important chapter of the book.

Homogenization and electromagnetic characterization of MMs have attracted a lot of attention of scientists since 2003. Many scientific groups studied MMs with WSD, and most often these MMs were effective media with resonant artificial magnetism. Unfortunately, the majority of reported results are disputable. If the reader of [1] agrees with the ideas of this book, this reader will recognize many publications on MMs erroneous. One of the most striking errors is persistent in the whole MM literature. In 2002–2004 this artifact was called *antiresonance of metamaterials* (see, e.g., [7]).

Antiresonance (anti-Lorentzian resonance) is a strange frequency dispersion of effective material parameters (EMPs) of the medium and a strange sign of their imaginary parts. Both dispersion and sign are not compatible with the requirements of basic physics. I think that such EMPs violating the basic physics are prohibited. If we obtained for an effectively continuous body a set of antiresonant material parameters, we made a mistake. If the body is not effectively continuous, the set of EMPs obtained after an assumption that the body is continuous is also wrong. In both cases these EMPs are equally meaningless.

The history of antiresonance started in 2002 from two simultaneously published works [8] and [9] where the so-called

Nicolson–Ross–Weir retrieval method was applied to MMs with resonant artificial magnetism. This method assumes the effective-medium slab to be uniformly continuous between its physical interfaces. Since it is simple, it turned out to be attractive for the community of metamaterial researchers. After these two publications, hundreds of MMs with artificial magnetism and/or bianisotropy were designed and in hundreds of scientific papers the Nicolson–Ross–Weir retrieval procedure was applied in order to retrieve the EMPs. As a result, in 2006 antiresonance was almost commonly recognized to be a new physical phenomenon inherent to MMs. However, in accordance with what I wrote in [1], the Nicolson–Ross–Weir procedure could not be applied to the media with resonant artificial magnetism and/or bianisotropy. In [1] it was pointed out that the resonance of the magnetic dipoles may perturb the continuity of the tangential components E_t and H_t of the mean electromagnetic field vectors E and H at the effective-medium interfaces. The resonance of any multipole except the electric dipole destroys the effective continuity of the interface. However, the failure of the primitive homogenization model does not obviously mean that the composite medium is effectively discrete. It only means that the primitive homogenization model must be revised. The necessity of the model revision was clearly pointed out in [1]. However, this revised model was developed later.

The revision of the quasi-static homogenization model was done in 2006 and I was inspired to do it by my friend Sergei Tretyakov (Helsinki University of Technology). In this revised model, the problem of the discontinuous interface was solved easily— the boundary jumps of E_t and H_t at the physical interfaces of the composite slab were mimicked by the so-called *transition layers*. This idea and some calculations confirming its consistency were reported in [10, 11]. Further, together with Sergei Tretyakov, I developed a modified retrieval procedure taking into account the peculiarity of the MM interface and successfully applied this algorithm to several finite-thickness MMs. We have seen that the retrieved EMPs experienced the resonance of the Lorentz type [12], whereas the standard retrieval resulted in antiresonance. Next, we have shown that the EMPs retrieved for a MM slab were exactly equal to EMPs previously calculated via the individual polarizabilities of the constitutive particles [13]. In 2007 we published a paper [14] in which these results were summarized.

This model was later developed in works [15–17] with the emphasis on the equivalence of the retrieved EMPs of different MM slabs and EMPs directly calculated for corresponding infinite MMs. This equivalence turned out to be an intrinsic property of a large group of regular MMs with electric and magnetic dipole responses that I have called *Bloch's metamaterial lattices*. Simultaneously, in works [18, 19] Mario Silveirinha has independently developed a similar homogenization model applicable to the same type of lattices. Silveirinha avoided the discussion of antiresonance and included the difficult case of so-called *impressed harmonic sources*. This study made Silveirinha's homogenization model applicable to the case when the sources can be embedded into an MM. My theory was restricted by a much simpler consideration of the lattice eigenmodes. In this part, our two homogenization models turned out to be completely coinciding. To see it, one may compare my work [17] and work [18]. However, in Silveirinha's model a closed-form analytical retrieval procedure is absent, because in order to solve the boundary problem, Silveirinha used *additional boundary conditions*, whereas I used a heuristic concept of transition layers.

These works served as a background for the scientific coordinating and support action ECONAM (Electromagnetic Characterization of Nanostructured Materials), in which I was appointed a scientific coordinator. This project started in 2008 as a part of the seventh scientific framework program of the European Commission (FP-7). The expert board and the governing board of ECONAM were united by the idea to improve the situation with the description of MMs in the scientific literature. In 2010 in collaboration with the teams of other on-going FP-7 projects related to metamaterials, the ECONAM team published a brochure [6] edited by an officer of European Commission, Anne de Baas. One of the purposes of this brochure was a concise explanation of two most frequent mistakes in the electromagnetic characterization of nanostructured MMs. One common mistake referred to the ellipsometric characterization of so-called *metasurfaces* [20].[a] The second common mistake referred to bulk MM and it was namely antiresonance.

In 2009 Filiberto Bilotti and I performed one of the main tasks of ECONAM—read all papers on MMs published in *Physical Review Letters*, the most popular physical journal, in 2002–2008. We analyzed the papers in which the EMPs of MMs were calculated or/and measured (78 works). Forty-one papers, in our opinion, supported by the expert board of the ECONAM consortium, were erroneous. Therefore, in [6] we pointed out this situation as a trouble for the future development of MMs. The brochure was later distributed among all European scientific groups specializing in MMs. Judging upon the literature, the reading of this book affected the metamaterial scientific community. At least, the number of papers containing the antiresonant EMPs has dropped noticeably since 2010.

In 2011–2012, a joint scientific group guided by two American scientists Chris Holloway and Edward Kuester published two papers [21, 22] on MMs with artificial magnetism. In these works, two finite-thickness transition layers were replaced with two sheets of surface electric and magnetic polarizations. Note that a similar work was independently published in 2011 by another team of coauthors [23], though it referred not to MMs but to mesoscopic media. Mesoscopic media are the media without a resonance of constituents but with an imperfect locality because their electromagnetic response depends on the sample thickness. The simplest example of a mesoscopic medium is a stack of optically thin dielectric layers with two alternating values of the permittivity—ε_1 and ε_2. The electromagnetic responses of samples with the even number of layers and those with their odd number are different. Although these media are effectively continuous and not internally resonant, it is not easy to properly homogenize them. In the present book, we do not consider the mesoscopic media and refer the reader to our paper [23]. In [23] and in works [21, 22], the boundary steps of the tangential mean fields E_t and H_t were described as effective sheets of electric (and, perhaps, magnetic) surface current at the slab interfaces. Surface susceptibilities describing these fictitious

---

[a]In this book we only briefly discuss metasurfaces concentrating on bulk composites.

currents were retrieved (together with the bulk EMPs of the effective medium sample) from the scattering matrix. This retrieval was done similarly to my retrieval in the model of transition layers. However, the boundary step of E_t and H_t is not spread over a fictitious transition layer but is referred to a physical interface at which it really occurs.

Next, in 2011 Andrea Alú, in work [24], revealed that the WSD can manifest in the antiresonance not only of wrongly retrieved EMPs but also of seemingly correctly retrieved EMPs of some infinite periodic lattices. My first-principle homogenization model developed in [14–17] turned out to be not self-consistent for some resonant lattices and needed a revision. This revision was done in works [24, 25]. It was shown that the continuity of the effective medium at the edges of the resonant stopband demands an introduction of an additional material parameter, similar to the *magnetoelectric coupling* parameter of a bianisotropic medium. Challenges of the resonant WSD were analyzed by several scientists. The effect revealed by Alú turned out to be significant for regular lattices with very small dissipative losses and very large resonance magnitude. If the array is not strictly regular or if its optical losses are substantial, the resonant stopband replaces by the band of resonant dissipation and the first-principle dynamic model keeps adequate [26, 27].

In view of all these advances, my book [1] has become obsolete. However, its content kept actual and I only needed to improve its text according to the new knowledge. This has been done in the present book, where the homogenization theory of metamaterials has been added. This book will hopefully serve to a better understanding of WSD and will help the specialists to correctly describe and characterize metamaterials.

To conclude the preface, I want to thank my friend Sergei Tretyakov, who inspired me in 2006 to search the exact reason for antiresonance in MMs (it turned out the misinterpretation of the so-called *Bloch impedance* of the MM slab treated by researchers as the *wave impedance*). In 2006–2007 Sergei led very useful discussions with me and contributed to my modified retrieval procedure. Later, he guided a collaborative research [27] that clarified the role of losses and aperiodicity for the applicability of the first-principle homogenization model. His contribution to the last chapter of this book is important.

I want to thank Antti Räisänen, who was the head of the Department of Radio Science and Engineering at Helsinki University of Technology, to which I moved in 2008 from St. Petersburg, for the favorable atmosphere in the department. I want to thank three most active members of the expert board of ECONAM: Filiberto Bilotti, Alexander Schuchinsky, and Alexey Vinogradov, who strongly contributed to the analysis of mistakes in the metamaterial literature. I want to thank the ECONAM project coordinator and the head of the governing board, Ari Sihvola, and the project manager, Vladimir Podlozny, who did their best for the successful dissemination of the project results. I want to thank my former bachelor, master, and PhD student, now my dear friend, Pavel Belov. First, his contribution to the proof of the generalized extinction theorem (see below) was principal. Second, he has persuaded me to write the present book.

Constantin Simovski

References

1. C. R. Simovski, *Weak Spatial Dispersion in Composite Media*, St. Petersburg, Polytechnica, 2003 (in Russian).
2. J. B. Pendry, Negative refraction makes a perfect lens, *Phys. Rev. Lett.* **85**, 3966–3969, 2000.
3. V. G. Veselago, The electrodynamics of substances with simultaneously negative values of ϵ and μ, *Sov. Phys. Uspekhi* **10**, 509–514, 1968.
4. F. Capolino, Editor, *Metamaterials Handbook, Vol. 1: Theory and Phenomena of Metamaterials*, CRC Press, London–NY, 2009. Chapter 2: Material parameters and field energy in reciprocal composite media.
5. L. D. Landau and E. M. Lifshitz, *Electrodynamics of Continuous Media*, Oxford University Press, Oxford, UK, 1982.
6. *Nanostructured Metamaterials*, A. de Baas, Editor-in-Chief, European Commission, Directorate General for Research, Publication Office of the European Union, B-1049 Brussels, 2010.

7. D. R. Smith and J. B. Pendry, Homogenization of metamaterials by field averaging, *JOSA B* **23**, 391–403, 2006.
8. D. R. Smith, S. Schultz, P. Markos and C. M. Soukoulis, Determination of effective permittivity and permeability of metamaterials from reflection and transmission coefficients, *Phys. Rev. B* **65**, 195104, 2002.
9. S. O'Brien and J. B. Pendry, Magnetic activity at infrared frequencies in structured metallic photonic crystals, *J. Phys.: Condens. Matter* **14**, 6383–6394, 2002.
10. C. R. Simovski, On the homogenization of artificial lattices, *Days on Diffraction'2006*, Steklov Institute, St. Petersburg, Russia, May 30–June 2, 2006, pp. 70–71.
11. C. Simovski, I. Kolmakov and S. Tretyakov, Approaches to the homogenization of periodical metamaterials, *Proc. MMET'06*, 11-th International Conference on Mathematical Methods in Electromagnetic Theory, Kharkov, Ukraine, June 26–29, 2006, pp. 41–44.
12. S. Tretyakov, C. Simovski and I. Kolmakov, Challenges in effective media modeling of artificial materials, *Third Workshop on Metamaterials and Special Materials for Electromagnetic Applications and TLC*, Rome, 30–31 March 2006, p. 26.
13. C. R. Simovski and S. A. Tretyakov, On effective material parameters of metamaterials, *Proc. 23-d Annual Review of Progress in Applied Computational Electromagnetics*, Verona, Italy, 19–23 March 2007, pp. 150–154.
14. C. R. Simovski and S. A. Tretyakov, Local constitutive parameters of metamaterials from an effective-medium perspective, *Phys. Rev. B* **75**, 195111(1–9), 2007.
15. C. R. Simovski, Bloch material parameters of magneto-dielectric metamaterials and the concept of Bloch lattices, *Metamaterials* **1**, 62–80, 2007.
16. C. R. Simovski, Analytical modelling of double-negative composites, *Metamaterials* **2**, 169–185, 2008.
17. C. R. Simovski, On material parameters of metamaterials (review), *Optics and Spectroscopy* **107**, 726–753, 2009.
18. M. G. Silveirinha, Metamaterial homogenization approach with application to the characterization of microstructured composites with negative parameters, *Phys. Rev. B* **75**, 115104, 2007.
19. M. G. Silveirinha, Poynting vector, heating rate, and stored energy in structured materials: A first-principles derivation, *Phys. Rev. B* **80**, 235120, 2009.

20. S. B. Glybovski, S. A. Tretyakov, P. A. Belov, Y. S. Kivshar and C. R. Simovski, Metasurfaces: From microwaves to visible, *Phys. Rep.* **634**, 1–72, 2016.
21. S. Kim, E. F. Kuester, C. L. Holloway, A. D. Sher and J. Baker-Jarvis, Boundary effects on the determination of metamaterial parameters from normal incidence reflection and transmission measurements, *IEEE Trans. Antennas Propag.* **59**, 2226–2240, 2011.
22. S. Kim, E. F. Kuester, C. L. Holloway, A. D. Sher and J. Baker-Jarvis, Effective material property extraction of a metamaterial by taking boundary effects into account at TE/TM polarized incidence, *Prog. Electromagn. Res. B* **36**, 1–33, 2012.
23. A. P. Vinogradov, A. I. Ignatov, A. M. Merzlikin, S. A. Tretyakov and C. R. Simovski, Additional effective-medium parameters for composite materials – excess surface currents, *Opt. Express* **19**, 6699–6704, 2011.
24. A. Alù, Restoring the physical meaning of metamaterial constitutive parameters, *Phys. Rev. B* **83**, 081102, 2011.
25. A. Alù, First-principles homogenization theory for periodic metamaterials, *Phys. Rev. B* **84**, 075153, 2011.
26. A. Andryieuski, S. Ha, A. A. Sukhorukov, Y. S. Kivshar and A. V. Lavrinenko, Unified approach for retrieval of effective parameters of metamaterials, *Proc. SPIE* **8070**, 807008, 2011.
27. P. Alitalo, A. E. Culhaoglu, C. R. Simovski and S. A. Tretyakov, Experimental study of anti-resonant behavior of material parameters in periodic and aperiodic composite materials, *J. Appl. Phys.* **113**, 224903, 2013.

Chapter 1

General Introduction to Weak Spatial Dispersion

1.1 Natural and Composite Media with Weak Spatial Dispersion

Modern optics and microwave technique cannot be imagined without composite media [28]. At the end of the 20th century, optical composites, including nanostructured ones (arrays of quantum dots, optical photonic crystals, nanostructured photovoltaic cells, etc.), found important applications in lasers, optical information and sensing systems, light energy harvesting, etc. This perspective allowed some authors to claim in the 1990s a start of the optical revolution in technique (see [29, 30]). In the microwave technique, composite media called artificial dielectrics and magnetics found applications as antireflecting shields (absorbers of the radar signal), in antenna radomes, in so-called *reflectarrays and transmitarrays* (passive devices, correcting the wave front of microwave radiation and shaping the pattern of both receiving and transmitting antennas), in so-called *Luneburg lenses*, performing the similar functionality, in polarization transformers for radars and radio telescopes, etc. (see [31–33]).

Composite Media with Weak Spatial Dispersion
Constantin Simovski
Copyright © 2018 Pan Stanford Publishing Pte. Ltd.
ISBN 978-981-4774-83-3 (Hardcover), 978-1-351-16624-9 (eBook)
www.panstanford.com

Composite media contain inclusions—constitutive elements of the composite—located in the host material called *matrix*. As a rule, this matrix is a transparent medium, whose role is to hold the inclusions. The matrix in the absolute majority of cases is a usual isotropic dielectric, whereas inclusions manifest a high optical contrast with it. There is a great variety of the sizes of constitutive inclusions. They can be negligibly small compared to the effective wavelength, e.g., molecules in molecular composites obtained by the molecular epitaxy or transferred into the host medium by the pressure of light in a powerful laser pulse. Artificial molecular media found applications in optics. For example, non-linearity of these molecular composites combined with the so-called *chirality* allows the efficient generation of the second harmonic of laser radiation [34].

Constitutive particles of a composite can have sizes comparable with or even larger than the effective wavelength λ_{eff} of the medium eigenmode. The most important class of such composites are so-called *photonic crystals*—regular arrays of inclusions that behave as effectively discrete media [35]. For many photonic crystals, the high optical contrast with the matrix is not necessary. For example, the inclusions can be voids in the dielectric matrix. Photonic crystals are mainly used in optics (see [35–37]) where they found numerous applications already toward the end of the 20th century.

In this book, we will not study photonic crystals. We will concentrate on the media which are effectively continuous for electromagnetic waves. For example, natural media in the visible light and impinged by electromagnetic radiation at lower frequencies are effectively continuous. Does their continuity mean that the sizes of their molecules are always so small compared to λ_{eff} that the electromagnetic field retardation effects in molecules are obviously negligible?

1.1.1 Chirality

No, it does not. If it was so, the continuity always implied the absence of spatial dispersion. However, there are molecules of complex shape that are though very small but not negligibly small compared to the wavelength. And the electromagnetic retardation in them results in an exciting property of the molecular medium. This property is already mentioned chirality. It is a special case of a more general phenomenon called *bianisotropy* or *magneto-electric coupling*. Chirality implies that the medium eigenmodes are circularly polarized and the medium response to the waves with clockwise and counter-clockwise polarizations is different. This phenomenon in the general case (for both reciprocal and non-reciprocal media) is called gyrotropy [5, 40]. Gyrotropy manifests in the difference of phase velocities of two circularly polarized eigenmodes (that is called *optical activity*) and in difference of their attenuation rate (that is called *dichroism*). The same effects—optical activity and dichroism—are observed in such natural materials as *ferrites*.

In ferrites, the gyrotropy arises at microwaves due to the asymmetry of the magnetic permeability tensor. In optically transparent ferrites, the gyrotropy arises also in the optical range—due to the asymmetry of the dielectric permittivity tensor. Ferrites are essentially non-reciprocal media. In order to distinguish optical activity and dichroism of non-reciprocal materials from these effects in reciprocal materials, many authors utilize the term gyrotropy only concerning the non-reciprocal media. Then the term chirality refers only to reciprocal media [38, 41, 42]. We follow the terminology of [40], where chirality is called a reciprocal implementation of gyrotropy.

Non-reciprocal gyrotropy arises in the dc magnetic field that breaks the spatial symmetry of an electromagnetic problem. Chirality arises due to the mirror asymmetry of molecules—scattering centers composing the effective medium. Geometric asymmetry of a chiral scatterer cannot destroy the reciprocity [5, 38, 40–42]. Sucrose and glucose molecules in water solutions, molecules of cholesteric liquid crystals, molecules of tartaric acid and DNA, proteins, etc., are chiral macromolecules. These macromolecules have helicoidal shape and have the maximal size in the range $a = 2 \cdots 25$ nm, whereas the visible range of wavelengths is $\lambda = [400, 800]$ nm. Chirality results from the molecule helicity. A molecule with higher helicity (more complex 3D shape) manifests higher impact of the retardation. Really, the path of the polarization current along the helix is much longer that the straight distance between the helix extremities. Therefore, for a helical molecule, the retardation effects are more pronounced than for a straight molecule of the same overall size.

Of course, the difference between λ and a in two orders of magnitude implies that the retardation effect is still very small. Respectively, the material parameter describing the chiral effect in the medium—chirality parameter—is small. Therefore, the polarization rotation in natural chiral media is measurable only in optically thick layer. It is not a resonant effect and is not accompanied by the practically measurable artificial magnetism.

Although the chirality parameter of natural chiral media is small, and the polarization rotation is not important in the scale of one wavelength, a practical chiral plate is 1 mm thick or thicker. It is sufficient so that to have the output polarization turned by 90° with respect to the input one [40]. Chirality of natural media found applications in many optical devices and serves an example of the importance of coherent microscopic effects for the effective medium. A weak microscopic effect when it is coherent has a strong impact to macroscopic wave processes [38–40, 43, 44].

Notice that chirality is not obviously related to optically small helicoidal molecules. Monocrystals of colored quartz also possess both optical effects referred to as chirality [38, 45]. Chiral effects arises due to the geometry of the so-called *superlattice* of quartz. A simple cubic lattice of a non-colored quartz in the case of a colored quartz is weakly screwed. The helical periodicity of the superlattice is comparable with the wavelengths of the visible range (400 − 800 nm). The helicoidal distortion of a cubic lattice is not detectable in the scale of one unit cell (0.1 nm).

The impact of the superlattice helicity can be described either by the magneto-electric coupling parameter [45] (and in this model quartz is a chiral medium) or, alternatively, as an inhomogeneous dielectric medium with helical periodicity of its permittivity $\bar{\bar{\varepsilon}}$. In this model, dielectric permittivity of the colored quartz is a uniaxial tensor with a constant axial component and transverse component periodically varying along the optical axis [43]. This is, probably, a more adequate description of quartz. Another example of a seemingly chiral medium without a microscopic chirality is a class of so-called *sculptured thin films* also called *columnar films*.

These films are molecular polymers with optically long (several micrometers) helicoidal molecules oriented normally to the film interfaces [145]. Sculptured thin films as well as colored quartz can be also described as a periodically inhomogeneous but effectively continuous dielectric [44]. Our book does not concern such media namely because their bianisotropy has no microscopic origin. Our theory refers to the composite or molecular media whose spatial dispersion originates from the response of constitutive particles. Particles of the medium in our theory are obviously optically small. Otherwise the medium cannot be effectively continuous. As to molecular media, our theory covers sugar solutions, cholesteric liquid crystals, chiral ferroelectrics, and other molecular media in which the constitutive particles are optically small and nevertheless manifest the chirality.

However, the most interesting situation for us is the case when the constitutive particles have sizes much larger than $\lambda/100$. Then the impact of the wave retardation is more pronounced and the effects of WSD such as chirality are manifested in layers whose thickness is not optically large. This is possible only in composite media.

As it was mentioned above, chirality is only one special case of bianisotropy. We are interested in understanding other types of bianisotropy, e.g., the so-called *Omega-type of magneto-electric coupling* (see below). Also, we aim to understand and describe the phenomenon of artificial magnetism. Both bianisotropy and artificial magnetism can be very strong if particles and inter-particle distances are not very small compared to the wavelength λ_eff in the effective medium. Practically, in composites with WSD the medium unit cell is larger than $(0.01 - 0.03)\lambda_\text{eff}$ (that is typical for natural chiral media) but must be not larger than $(0.2 - 0.3)\lambda_\text{eff}$, otherwise the spatial dispersion becomes strong, and the medium cannot be described as effectively continuous.

Optical density of a medium formulated in terms of a sufficiently small ratio $a/\lambda_\text{eff} < 0.2 - 0.3$ is the most important condition of the effective continuity of the composite. What does it mean— effective continuity—for the electromagnetic response of a medium? In the isotropic case, the effective continuity implies that the incident plane wave basically excites an only eigenmode of the effective medium called a refracted plane wave. Other excited eigenmodes are negligibly small. Also, an effectively continuous slab illuminated by a plane wave does not produce scattered waves outside it. For anisotropic or bianisotropic effectively continuous media, the incident plane wave may excite one ordinary wave and one (in uniaxial media) or two (in the most general case of biaxial anisotropic media) extraordinary waves. Also, no extra scattered waves are allowed in front of or behind the medium slab. Besides these evident implications of the continuity, there are other more fine implications that will be discussed below.

1.1.2 Artificial Magnetism

The most interesting case of a composite is that its constitutive particles experience the resonance. Then bianisotropy and/or artificial magnetism can be so strong that the electromagnetic response of a composite material breaks the standard insight elaborated for continuous media in the 19th and 20th centuries. Such composite media may be effectively continuous but acquire the extreme properties at the magnetic or bianisotropic resonance that earlier considered impossible for continuous media (such as violation of Maxwell's boundary conditions at the interfaces). These composites can be referred to as MMs according to their modern definition.

Some composites become effectively discrete at the resonance of their constituents. If they are regular lattices, they are referred to as photonic crystals (in the corresponding frequency band), if they are internally random, they are referred to as scattering media, that can be either turbid or completely opaque. The phenomenon of WSD does not refer to photonic crystals or scattering media. The spatial dispersion in them is strong. Below we will discuss the difference between weak and strong spatial dispersion.

Probably, the most important manifestations of the resonant WSD refer to nanostructured MMs operating in the optical range. Especially, the resonant artificial magnetism is an exotic phenomenon for optical frequencies, because it these frequencies the natural magnetism is not observed. Optical composites with artificial magnetism have been created only recently. However, artificial magnetism at rather low frequencies has been known since the experiments of Michael Faraday on the electromagnetic induction. In the 1940s, composite media with artificial magnetism—mixtures of metal inclusions shaped as split rings—were engineered (see [46, 47]). Artificial magnetism in the effective media arises due to curls of the polarization current, i.e., due to the Faraday effect in the embedded rings. Recall that according to Faraday, a time-varying magnetic field induces a current in a wire loop. Therefore, in a medium formed by an array of loops, the wave propagation is accompanied by the excitation of the curl currents. These microscopic curls after a proper homogenization result in the mean magnetization of the effective medium by the wave field.

This effect is very weak and results in the certain artificial diamagnetism if the rings are solid. A sub-millimeter split in a wire ring performed of a mm-thick Cu or Al wire behaves as a lumped capacitor. Usually, there is a tiny dielectric insertion in the split preventing the ohmic contact across it. This capacitive loading offers a paramagnetic resonance to the ring which is inductive itself. Namely this resonance was achieved at microwaves in the 1940s (see e.g. [46, 47]). The purpose of this *split-ring resonator* (SRR) was to be a constituent of an isotropic artificial medium whose effective permeability would attain at the operation frequency the value of the order of $2.5 \cdots 3$—as high as that of the effective permittivity of this composite [47].

When the relative permittivity ε and permeability μ of an isotropic medium coincide, the medium wave impedance becomes equal to that of free space. Then the normal plane-wave reflectance of the medium vanishes. Fitting the frequency range in which $\varepsilon \approx \mu$ to the operation frequency of an adversary naval radars one prevents the radio location of a military ship whose metal body is covered by such a magneto-dielectric shell. The source wave is practically not reflected from such a ship, it transmits into its shell and dissipates there. This idea was called the *matched absorber* and it refers to the early stage of so-called *stealth technologies* [31, 46].

Of course, the matched absorber is not an only method for the suppression of the microwave reflection from large metal objects. A popular antireflective coating of a big metallic object is the so-called *Dallenbach's absorber*, a metal-backed dielectric layer with rather low (though sufficient) optical losses and properly chosen thickness. This thickness is the odd multiple of a quarter of the effective wavelength in the medium $\lambda_{\text{eff}}/4$, i.e., $(2N+1)\lambda_{\text{eff}}/4$, where $N = 0, 1, 2, \ldots$. This absorber utilizes the Fabry-Perot resonance making the absorption in the dielectric coating nearly total. For the same purpose one may use a lossless dielectric coating of the same thickness covered by a thin absorbing sheet. This structure is called the *Salisbury's absorber*. Both Dallenbach's and Salisbury's absorbers were often used when the struggle against narrow-band radars was actual. When the radars with sweeping frequency were created, the multilayer broadband extensions of Dallenbach's and Salisbury's absorbers were developed [31, 46]. However, still matched absorbers kept actual. During the exploitation period, the abrasion changes the thickness of the cover deteriorating its operation. Media with $\varepsilon \approx \mu$ are advantageous because their matching to free space does not depend on the thickness. This is so if the thickness exceeds the field penetration depth [31, 46]. This is why matched absorbers are still used [31, 46] (see [50]), though nowadays, condition $\varepsilon \approx \mu$ is achieved in a broad frequency band. This is possible using the mixtures of submicron iron particles (nanomagnets) in a silicone rubber host.

As to microwave composites with *artificial* magnetism (and/or bianisotropy), they are fabricated, nowadays, using the planar technology and stacking the composite monolayers. For it one uses printed circuit boards allowing the planar metallic particles. The stacks of patterned printed-circuit boards can be pierced by metal vias of mm or slightly sub-mm thickness. In this way one prepares 3D complex-shape particles from planar ones [48]. Multilayers of 3D complex shaped particles can be also fabricated alternatively—using the so-called *LTCC technology* [49].

1.1.3 About Non-Local Description of Media with Chirality and Artificial Magnetism

Both magnetic permeability of composite media based on SRRs and chirality of natural media of helicoidal molecules result from the general phenomenon called spatial dispersion (SD). And in both cases this SD is called weak spatial dispersion. WSD does not mean that the effect is weak. It means that these media, in spite of SD in them, can be considered as effectively continuous materials. Weakness of SD means that the phenomenon of SD does not destroy the effective continuity. Strong SD destroys the continuity of the effective medium and it cannot be described via a set of EMPs unique for all waves propagating in the medium. Description of media by a set of such EMPs is called *local homogenization model*.

In the local model, media with WSD can be described if besides the effective permittivity, a special material parameter, responsible for the artificial magnetism, is introduced (this parameter is called permeability) and a special material parameter, responsible for the bianisotropy is introduced (this parameter is called coefficient of magneto-electric coupling).

There is a possibility to describe the same media as a discrete one. This model does not take into account that the medium behaves as an effectively continuous one. It takes into account the discreteness of the original array and describes its electromagnetic response to a given eigenwave. Due to absence of the natural magnetism, one may describe the medium with WSD in terms of an only dyad called non-local permittivity [5]. Non-locality means that this permittivity tensor depends on the wave vector—for the given frequency depending on the direction of the wave propagation. Such a tensor describes the polarization impact of the wave and may be useful for the analysis of eigenwaves.

However, using only the non-local permittivity, we cannot find the amplitude and phase of an eigenmode excited in a spatially bounded sample of the medium. To solve the boundary problems utilizing the non-local permittivity, one has to deduce also the so-called *additional boundary conditions* [51].

Additional boundary conditions (ABCs) are rarely deducible analytically for realistic composite media. Even if they can be derived in a closed form, this approach to the solution of boundary problems is an unnecessary complication for a medium that is effectively continuous. Therefore, our theory of WSD aims namely the description of the media in terms of local material parameters—those independent on the wave vector **q**.

It is worth to notice that the artificial magnetism is not always a simple excitation of a magnetic dipole moment in a constitutive particle. Not all MMs with artificial magnetism are composed by SRRs or other similar scatterers whose magnetic response is simply dipolar. Artificial magnetism is often (especially in optical MMs) accompanied by the strong excitation of higher multipoles, and among them electric multipoles can be present. MMs with a multipole response may be challenging for homogenization. Sometimes, they are and at the same frequency, effectively continuous for waves propagating for a certain sheer of the propagation directions and effectively discontinuous for other directions. This issue will be discussed below in details through the example of a MM composed by dimers. For such media, it is reasonable to find, for every frequency, a sheer of propagation directions for which the SD in a medium with artificial magnetism is weak, and abstain of applying our theory for other angles where SD is strong.

Now, let us discuss the interplay of the strong and weak SD over the frequency axis. The concept of WSD makes sense only within some frequency ranges. Beyond these ranges, a composite is either a simple dielectric or, on the contrary, an effectively discrete medium. At sufficiently low frequencies starting from the zero one, the composite is always a simple dielectric. The effects of WSD—bianisotropy and artificial magnetism—are noticeable only at sufficiently high frequencies.

To conclude this subsection let us discuss the difference between the natural and artificial types of magnetism. Natural magnetic moment (e.g., that of a domain in a ferrite or a ferromagnetic) results not from the electromagnetic induction, it intrinsically exists on the microscopic level, at least in present the constant magnetic field. Therefore, the corresponding magnetic permeability of the medium does not degenerate in statics into unity. Magnetic domains of natural magneto-dielectric media (transparent like ferrites or impenetrable like ferromagnetic media) are very small compared to the wavelength at all frequencies where they interact with the electromagnetic fields. Therefore, the additional magnetization of a naturally magnetic medium produced by the electromagnetic wave results not from the retardation effects, and no classical curl currents can be attributed to magnetic domains. The action of the magnetic field of a wave in such media is not reducible to that of the spatially varying electric field. This is another difference of the artificial magnetism from its natural analogue besides of the medium reciprocity.

1.2 Homogenization

1.2.1 Homogenization in General

In the present introductory chapter, we appeal to an intuitive—phenomenological—approach to the macroscopic and microscopic polarization, used in books [5, 41, 42, 52, 55, 56]. It is enough to understand the concepts of homogenization, continuity, and locality.

Homogenization of a medium is its macroscopic description in terms of a few effective material parameters. These EMPs should be unique for all medium eigenmodes that can propagate at a given frequency. They can be functions of the frequency ω but should not depend on the wave vector $\mathbf{q}$. One may introduce different sets of material parameters which would be functions of $\mathbf{q}$, and even operators acting on the field phasors, but this description of the medium in the terminology of the present book is not called homogenization. In this book, homogenization refers only to media which can be considered as effectively continuous. Since we have performed the homogenization, we can forget that the original structure consists of particles. Now we deal with polarization currents and charges which are continuous functions of coordinates. Our EMPs must be compatible with the idea of continuous media and cannot violate causality and passivity principles imposed for such media.

The final result of a homogenization model is a set of relations between the response of an individual inclusion to the electromagnetic field acting on it and the EMPs of the effective medium formed by all these inclusions. This set of relation is called *mixing rules*. Mixing rules should be derived analytically, otherwise the homogenization model is hardly useful.

Before the derivation of these relations, we have to establish the *material equations* (MEs) into these EMPs would enter. MEs should be written analytically in a closed form. The derivation of MEs is the key point of the homogenization model.

As to the response of an individual particle to the local field, it can be found separately—either analytically evaluated or simulated numerically. For complex-shaped particles, full-wave numerical simulations or experimental validation is necessary even if an approximate analytical model is known.

Once the homogenization model for a given array of particles is built, one starts from the calculation of microscopic parameters—their analytical estimations and numerical simulations and finishes by EMPs. This is the direct application of the homogenization model. There is also an inverse application of the homogenization model called retrieval of EMPs. The retrieval procedures will be also discussed in this book.

Now, let us discuss some critical points of homogenization. Imagine that we have calculated the EMPs of a medium and they turned out to be prohibited for continuous media. How to understand the reason for this result? Is it so because the medium is effectively discrete and cannot be homogenized? Or is it a simple mistake?

First, we have to enquire whether there is evidence of the medium discreteness. As it was already mentioned, this discreteness for regular media implies the scattered waves (also called Fraunhofer diffraction side-lobes) and/or additional waves excited in the medium slab. Here, we add another evidence of the electromagnetic discreteness in the internally periodic media. This is the Bragg phenomenon—prohibited propagation of a wave at a given frequency in a given direction. The prohibition is not the same as the wave decay—the wave can decay due to dissipation or scattering losses. The prohibition means the destructive interference of partial waves produced by adjacent crystal planes in the direction of propagation and constructive interference of waves partially reflected by crystal planes. As a result, an incident wave exciting such a prohibited eigenmode is totally reflected.

The Bragg phenomenon was discovered a century ago in the experiments with X-rays in dielectric crystals. In the optical range, this phenomenon is inherent to the photonic crystals. At millimeter waves and microwaves it is observed in low-frequency analogues of photonic crystals, called *electromagnetic crystals* or *electromagnetic band-gap structures*. Also, the Bragg phenomenon is observed for the so-called *de Broglie waves*—those of the electron wave function—in solid semiconductors.

Over the frequency axis, the Bragg phenomenon starts at a frequency where the distance between the crystal planes equals to one half of the effective wavelengths. For lattices with substantial sizes of the constitutive particles the Bragg phenomenon forbids the propagation of the eigenmode in a range of frequencies with nonzero bandwidths. This range is called the Bragg stopband. If a slab of the effective medium impinged by a plane wave manifests the Bragg stopband, it is effectively discrete even in absence of additional waves and scattering maxima.

In absence of all these features of discreteness, a regular medium still cannot be described as effectively continuous if we study a broad sheer of propagation angles and the medium manifests the unusual type of the electromagnetic dispersion, impossible in continuous materials. Here, we mean the so-called *dispersion surfaces* [53], also called *Fresnel's wave surfaces* (see [5]), and *isofrequency surfaces* (see [42]). For brevity, in this book we will use the term *isofrequency*.

Isofrequencies are solutions of the dispersion problem of an infinite periodic structure for given frequencies. They represent constant-frequency plots in the axes q_x, q_y, q_z—coordinates of the so-called *reciprocal space*, where **q** is the lattice eigenmode wave vector. Also, 2D constant-frequency contours—in the planes $(q_x - q_y)$, $(q_x - q_z)$, and $(q_y - q_z)$—are called isofrequencies. Of course, isofrequencies make sense only for propagating waves, evanescent eigenmodes are beyond consideration.

Classical electrodynamics of continuous media with positive EMPs [5, 52, 53] states that isofrequencies can be only ellipsoids, for isotropic media—spheres. Nowadays, it is known that a material tensor of an anisotropic medium can have components of different signs. In this case the isofrequency has hyperbolic shape [70] and such artificial media, called hyperbolic materials or hyperbolic media refer to the class of MMs[a]. So, continuous media have either elliptic or hyperbolic isofrequencies. If an isofrequency of a regular

---

[a]In fact, an isofrequency of a cold magnetized plasma is also a hyperboloid that has been known since the 1950s (see [71]).

medium is not elliptic or hyperbolic the medium is, strictly speaking, effectively discontinuous at this frequency.

However, as it was already mentioned, there is a possibility to speak on effective continuity of such media for a certain sheer of propagation angles. Assume that the isofrequency of our material is a cube. Cube is not the special case of an ellipsoid or hyperboloid, and, strictly speaking, the medium manifests at this frequency a strong spatial dispersion. However, let us restrict our analysis by such propagation directions that the wave vector $\mathbf{q}$ belongs to one face of thus cube. For these propagation directions our material behaves as an effectively continuous medium with ultimate anisotropy. Really, a face of a cube is a piece of an infinite plane, that is the limit case of a hyperboloid! Propagating waves in the corresponding sheer of angles do not feel that the medium is effectively discrete. Therefore, EMPs calculated or retrieved for this special case must be physically sound and must satisfy to the necessary limitations established for continuous media.

We have specified all features of the electromagnetic continuity. If our EMPs turned out to be non-physical for the case when the medium behaves as an effectively continuous one, we made mistakes. For example, we used a wrong homogenization model.

1.2.2 About Bulk Homogenization

A reader studying the electrodynamics of continuous media from classical tutorials and monographs [5, 41, 42, 52–55] or even from recent books such as [38, 56, 63] often meets the terms *macroscopic* and *microscopic* without a mathematically strict definition of them. Of course, it is clear without strict definitions, that a wave process in a medium is macroscopic, whereas a polarization and/or magnetization of a medium particle is a microscopic process. Respectively, the electric or magnetic polarization of the medium elementary volume (electric or magnetic dipole moments per unit volume) is macroscopic polarization. The last one is described by EMPs, called medium susceptibilities, whereas the microscopic polarization is described by the microscopic parameters called the electric and magnetic dipole polarizabilities. However, this intuitive notion does not clarify the transition from the particle polarizability to the medium EMPs.

It is usually thought that macroscopic polarization of a medium is a simple smoothing/spreading of a microscopic one. First, it is not always so, and we will see in this book that for MMs a correct smoothing is not so simple. Second, even if the averaging can be a simple smoothing, it is not so simple to calculate EMPs via particle polarizailities. EMPs describe the response of the medium unit volume to the mean (macroscopic) field. This field results from the averaging of the *true field*, also called microscopic field. True field is created by the external sources and by all scattering particles (for composites—also by a continuous polarization of the host medium). Smoothing of the true field results in the mean field, and we have to find the response of the medium unit cell to this mean field.

However, the microscopic polarization of a particle is performed not by the true field. A particle cannot polarize by itself, it is polarized by the *local field*—that created by all other sources and scatterers around the particle. It is difficult to homogenize the bulk medium if we do not know how the local field is related with the mean field. Therefore, even a simplest quasi-static homogenization model is not a straightforward procedure. One must find and use a relation between the local and the mean field.

In books [5, 41, 52, 55], this part of the theory is omitted as such—media with given macroscopic parameters are assumed to be continuous from the start, their molecular/atomic structure is not analyzed. In books [42, 53, 54, 56], a quasi-static model of an effectively continuous medium is present. However, even in these books the homogenization theory is explained insufficiently. Here we do not imply composite media. Even the homogenization model of natural media in these books is insufficient. The main shortage is the absence of the interface homogenization.

1.2.3 About Homogenization of Interfaces

In classical books on electrodynamics of media, the reader can find the homogenization model developed by H. A. Lorentz. This is a simple and elegant quasi-static model. However, it is applicable only to an infinite array of dipole particles. Even if we admit from the start that our particles are polarized namely as electric dipoles and that the frequencies of our interest are sufficiently low, this model still loses the validity near the interface of a spatially bounded array. This model does not answer the key question: Can the layers of molecules located near the interface be described as an effectively continuous medium. In other words, it remains unproved that the medium interface is continuous.

Assume that the boundary of a medium is not continuous—the effectively continuity refers only to the bulk part of the molecular array. Then the incident wave must penetrate through the interface into the array. Well, deeply in the bulk the array becomes a continuous medium and can be described by the effective permittivity. Therefore, in the bulk the incident wave must be substituted by the refracted wave. However, in this case the Fresnel formulas for the wave reflection and transmission of a semi-infinite medium are not applicable because the incident wave is reflected not from the physical interface of a medium slab. Then how can we calculate the reflection and transmission coefficients?

So, homogenization of a bulk medium—the theory by H. A. Lorentz—does not allow us to analytically solve boundary problems. We need a valid model of homogenization for the medium interface.

Here, we may recall that physics is an experimental science and tell that numerous experiments have shown—the Fresnel formulas work very well for the absolute majority of natural media up to the ultraviolet frequency range. This means that the natural medium interface is effectively continuous. Although it remains unclear why it is continuous, we may postulate this for natural media.

However, after reading books [5, 41, 42, 52–56] and many other books on the electrodynamics of media, we learn nothing on the effective continuity of the *composite* medium interface. Will Fresnel's formulas be applicable for media whose particles are smaller than the wavelength only by one order of magnitude? Below we will answer this question. And the answer will be different for media with resonant artificial magnetism/resonant bianisotropy and without these resonances.

To conclude this subsection, we have to mention an only classical book [64] where the importance of the medium interface problem for the homogenization is properly explained. In this book, it is stated that the concept of homogeneous interface called *extinction principle* is a prerequisite of the application of any bulk homogenization model to a spatially bounded medium. However, even this book is insufficient for the understanding of this problem because the proof of the extinction principle in [64] is not self-consistent (see below). Probably, [65] is an only more-or-less known book where a valid proof of the extinction principle is presented.

1.2.4 How to Apply the Homogenization Model

The direct application of the homogenization model starts from the calculation of the particle polarizabilities. One needs to calculate the individual particle response to the electric external (local) field and its response to the magnetic external field. In the analytical model of the particle it is a straightforward procedure. But how can it be done in numerical simulations? And how can these polarizabilities be retrieved experimentally?

The response of a particle to the local electric field can be simulated by centering the particle exactly at the maximum of a standing wave. This maximum is the node for the external magnetic field **H**, and the impact of the local magnetic field is excluded in the simulated particle response. Locating the single particle at the node of the standing wave we obtain the response to the local magnetic field. These responses can be obtained also experimentally using two waves of the same amplitude with opposite propagation directions, e.g., in a resonator.

In simulations, the particle response is presented by the distribution of the polarization current over the particle volume. This distribution allows us to calculate either spherical or Cartesian multipoles induced by the local electric and magnetic fields. In this book we will use the Cartesian multipoles. They are more relevant for complex-shape particles and, nevertheless, applicable to spherical particles as well. From numerical simulations, we may find the multipole polarizabilities relating the induced Cartesian multipoles with the given local fields.

Applying the theory presented below, one may calculate EMPs through these polarizabilities. These EMPs must work for finite samples of the medium—at least for medium slabs infinitely extended in a plane. The obtained EMPs should correctly (with acceptable accuracy) predict the scattering matrix elements of a slab—its reflection R and transmission T coefficients—and also correctly predict the mean field inside the slab—the field formed by the refraction of the incident wave and internal reflections in the slab. Vice versa, our theory, which pretends to be self-consistent, must allow us to retrieve EMPs from measured or simulated R and T coefficients and using our theory to restore microscopic parameters—multipole polarizabilities of a single particle. Of course, the retrieved parameters must satisfy the Kramers–Kronig conditions. The role of these conditions was briefly explained in Preface and below we discuss them in more details.

Notice that full-wave numerical studies of a composite slab with regular internal geometry allow us both direct and inverse application of the theory. Using full-wave simulations for a regular finite-thickness slab and for an infinite lattice with so-called Bloch's boundary conditions for a unit cell we may compare the retrieved EMPs with those simulated for an infinite lattice. Normally, these two sets of parameters coincide. Then such lattices (at frequencies where they are optically dense) are media with WSD.[b] Next, these EMPs should be compared with those resulting from the particle polarizabilities and analytical formulas expressing our EMPs through these polarizabilities. If these sets of EMPs coincide with an acceptable accuracy our theory is fully consistent and working.

1.2.5 About Homogenization of Media with Weak Spatial Dispersion

If you do not want to go throughout a body of papers and prefer to familiarize with a homogenization model suitable for media with WSD in a book, you will find only two books [38, 63] comprising a detailed review of such homogenization models. Unfortunately, both these models are incomplete and do not suggest any practical algorithms suitable for finite bodies.

---

[b]Below we will discuss this class of optically dense lattices—the so-called *Bloch lattices*. Finite-thickness Bloch lattices can be homogenized even if they lose the effective continuity at the interfaces. We will see that the continuity of the lattice interfaces is lost if the constituents of the lattice experience the magnetic resonance.

A homogenization model suggested in [63] is called *two-scale dynamic homogenization*. It makes sense for condensed composites with artificial magnetism and bianisotropy. Below, we will briefly review this model and see that it is only applicable to an infinite bulk medium. The homogenization of the interface in this model is an open question. In [63], it is pointed out that this model is hardly compatible with Maxwell's boundary conditions and requires their revision. This revision was not done in [63], the author only promised to do it in future. A quasi-static homogenization model presented in [38] is similarly incomplete. It refers to infinitely extended bianisotropic media, does not consider the problem of the interface, and ignores the artificial magnetism. Therefore, the readers eager to understand how composite media with WSD can be homogenized have to browse journal and conference papers.

Now, let us briefly discuss the importance of homogenization for media with WSD. Homogenization gives a physical insight of the operation of an effectively continuous array of particles. This is instructive and may even result in a scientific breakthrough. For example, works [72] and [73] demonstrating the negative refraction in a doubly negative MM gave a powerful pulse to the development of MMs namely because these arrays were homogenized and manifested an exciting combination of the negative permittivity with negative permeability.

As to practical applications, and especially, engineering, homogenization is important for the synthesis of composites. To design a composite medium via full-wave numerical simulation of an array of particles is hardly fruitful. In this way the researcher will occupy computational resources for a long time without understanding what should be changed in the initial array in order to achieve its optimal operation. Homogenization simplifies this problem considerably. It gives the description of an array in a condensed form, through a few EMPs whose values are predicted with at least qualitative accuracy in a rather broad frequency range.

Of course, a certain error is unavoidable when applying the homogenization model to a practical array. However, a model of a continuous effective medium is a relevant estimate that drastically saves the resources and time needed for the synthesis. After a preliminary analytical design with approximately estimated optimal design parameters, a final numerical optimization would require much less full-wave numerical simulations than a blind numerical optimization. Also, a correct homogenization model grants a better understanding of the needed design.

Of course, the homogenization model is useful only if it is correct. An incorrect homogenization model is harmful; it will only disorient researchers, will result in wasted resources, and even may compromise the whole concept of WSD in the eyes of readers.

1.3 Weak Spatial Dispersion versus Strong Spatial Dispersion

Spatial dispersion can be defined for electromagnetically linear media without natural magnetism in two equivalent ways. The first way is the identification of SD with the *non-locality* of the medium response to the macroscopic (mean) electromagnetic field at a given frequency. The second one is based on the introduction of the effective permittivity that depends on both frequency and wave vector. The first way seems to be more physically relevant. Here we start from it and show the mathematical equivalence of the two approaches in the end of this section.

Non-locality means that the polarization response of the medium at the observation point $\mathbf{r}$ is determined by the macroscopic (mean) electromagnetic field not only at this point but in a certain spatial domain around it. In the most general form the polarization current density $\mathbf{J}$ at any observation point $\mathbf{r}$ is the response of the effective medium to the mean field $\mathbf{E}$ distributed in the whole space (volume V_∞). In the linear electrodynamics, it is a convolution of the mean field with a certain kernel K. In the general case, this kernel is a tensor. For the sake of simplicity, we omit in this section the tensor notations and write

$$\mathbf{J}(\mathbf{r}) = \int_{V_\infty} K(\mathbf{r} - \mathbf{r}')\mathbf{E}(\mathbf{r}')\, d^3\mathbf{r}'. \tag{1.1}$$

Magnetic mean field vector $\mathbf{H}$ or $\mathbf{B}$ in the right-hand side of (1.1) is absent because there is no natural magnetism. In the case of the natural magnetic media, there is a physical difference between magnetic tension vectors **H** and magnetic flux vector **B**. Physically it is so because the microscopic magnetization is originally present in such a medium even in statics. Time-varying fields only orient the magnetic moments; they do not induce them. Meanwhile, artificial magnetism results from vortex polarization currents in particles induced in particles by the time-varying fields. Then the action of the magnetic field in (1.1) is automatically taken into account. Really, if we write instead of (1.1) a similarly more general relation

$$\mathbf{J}(\mathbf{r}) = \int_{V_\infty} [K_{\text{el.}}(\mathbf{r}-\mathbf{r}')\mathbf{E}(\mathbf{r}') + K_{\text{mag.}}(\mathbf{r}-\mathbf{r}')\mathbf{B}(\mathbf{r}')] \, d^3\mathbf{r}',$$

the second term in the right-hand side can be merged with the first one because $\mathbf{B} = j(\omega)^{-1} \nabla \times \mathbf{E}$. Then we have for K in (1.1) a relation taking into account the separate medium responses to the electric and magnetic fields:

$$K(\mathbf{R}) = K_{\text{el.}}(\mathbf{R}) + \frac{j}{\omega} \nabla \times K_{\text{mag.}}(\mathbf{R}).$$

Thus, polarization currents (in an effectively continuous medium without natural magnetism)—both laminar and vortex components—are fully determined by spatially varying electric field. This is the rather general result, valid for reciprocal bianisotropic media and media with artificial magnetism.

Strictly speaking, Eq. (1.1) implies the integration over the whole space—it does not matter for finite or infinite arrays of particles. However, in the case of the medium with WSD in order to find the polarization current at the observation point **r** it is enough to perform the integration over an optically small volume V centered at **r**. In Chapter 4, we will accurately obtain this drastic reduction of the effective response volume as the necessary and sufficient condition of the medium effective continuity. Here we explain it qualitatively.

A macroscopic polarization current originates from the microscopic polarization of the particles located in the near vicinity of the observation point. If the medium is effectively continuous, it is optically dense. Therefore, the electromagnetic interaction of particles is performed by near fields. Calculating the microscopic polarization currents around the observation point and averaging them in order to find **J**(**r**), we may neglect the influence of the particles located at the wave distances from **r**. So, the mean polarization current **J(r)** is determined by the fields distributed in a rather small volume V centered by the observation point. The shape of this volume will be clarified below. Now we only need to understand that the size of the effective volume V replacing V_∞ in (1.1) for effectively continuous media is much smaller than λ_{eff}.

Since the volume V is optically small, the variation of the mean electric field over it is not drastic. Therefore, expanding **E** inside it into the Taylor series we obtain from (1.1):

$$\mathbf{J}(\mathbf{r}) = \mathbf{E}(\mathbf{r}) \int_V K(\mathbf{r}-\mathbf{r}') d^3\mathbf{r}' + \nabla \mathbf{E}(\mathbf{r}) \int_V K(\mathbf{r}-\mathbf{r}') r' d^3\mathbf{r}'$$
$$+ \frac{1}{2} \nabla\nabla \mathbf{E}(\mathbf{r}) \int_V K(\mathbf{r}-\mathbf{r}')(r')^2 d^3\mathbf{r}' + \cdots \quad (1.2)$$

Here, we have noticed that **E(r)** and its spatial derivatives $\nabla \mathbf{E}(\mathbf{r})$, $\nabla\nabla \mathbf{E}(\mathbf{r})$, etc., are taken at point **r**, i.e., do not depend on the integration point **r**'. Therefore, we have transposed **E(r)** out from the integrals. We may consider the second term in the right-hand side of (1.2) as that of the first order of smallness, the third term as that of the second order of smallness, etc. For an effectively continuous medium the mean polarization current density **J(r)** yields to a rapidly converging series of spatial derivatives of the mean field.

Notice that the operator form $\nabla \mathbf{E}$ contains both scalar product $(\nabla \cdot \mathbf{E} \equiv \text{div}\mathbf{E})$ and vector product $(\nabla \times \mathbf{E} \equiv \text{rot}\mathbf{E})$ of the del (nabla) operator to the mean electric field. The vector product in the second term in the right-hand side of (1.2) corresponds to the *bianisotropy*. Bianisotropy means that the magnetic field (curl of **E**) brings the contribution into the electric polarization current **J**. One may guess that in this case the locally uniform (but time-varying) electric field brings the contribution into the magnetic moment density of the medium. This guess will be confirmed below. As to artificial magnetism, it is not so easy to see it hidden in the third term in the right-hand side of (1.2). Below we will discuss it with all needed details.

Noticing that the integrals in Eq. (1.2) do not depend on the electromagnetic field and only depend on the medium response function, we may rewrite this equation in the form

$$\mathbf{J}(\mathbf{r}) = j\omega\kappa \mathbf{E}(\mathbf{r}) + j\omega\kappa' \nabla \mathbf{E}(\mathbf{r}) + j\omega\kappa'' \nabla\nabla \mathbf{E}(\mathbf{r}) + \cdots. \quad (1.3)$$

Here κ, κ', $\cdots$ are EMPs which can be called medium susceptibilities (why we share out the factor $j\omega$ will be clear from the strict consideration presented below). If all the terms in the right-hand side of (1.3) except the first (zero-order) term are negligible the SD is absent. However, the medium is still effectively continuous if the series converges. WSD is (by its classical definition in books [5, 42, 54, 75, 76]) the case when the second and third terms in (1.3) are not negligible, whereas all the omitted terms of the series are negligible.

If the effective-medium response volume V is optically large, the series (1.3) does not converge. This is the case when SD in the effective medium is strong, and the medium is not effectively continuous. A microscopically discrete nature of a medium described as an effectively continuous one can be essential for the electromagnetic interaction of particles located inside the volume V. However, microscopic discreteness as such is not harmful for our model. It only means that function K may be a step-wise function inside this volume, whereas $K = 0$ for integrations point $\mathbf{r}'$ located beyond this volume. Once we have calculated the EMPs entering (1.3), we may ignore the medium physical discreteness—it is electromagnetically continuous.

The situation for media with strong SD is opposite. Here the near fields do not dominate in the local field over the wave fields of the distant sources simply because the distance from the reference particle to the adjacent ones is also of the order of the effective wavelength. At these distances the near fields decay. The electromagnetic interaction of particles is carried out by wave fields and is governed by their interference. For a photonic crystal, a regular lattice of particles whose unit cell is optically substantial this interference is coherent.

In principle, the mean fields and, respectively, the set of EMPs can be introduced for photonic crystals. However, these parameters are rarely helpful in the solution of practical electromagnetic problems for photonic crystals. This is so because the set of EMPs of a photonic crystal is non-local—dependent on the wave vector $\mathbf{q}$. This dependence is strong, it can be even resonant and not reducible to the linear or quadratic function. This is called strong SD.

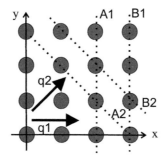

Figure 1.1 Illustration to the concept of strong spatial dispersion for a regular lattice. The local field comprises the sum of partial plane waves radiated by polarized crystal planes of the lattice along the wave path. At the same frequency the interference of partial plane waves produced by two adjacent crystal planes A_1 and B_1 (the case of the wave vector q_1) is evidently different from the interference of the plane waves produced by two adjacent crystal planes A_2 and B_2 (the case q_2). Therefore, the medium response is different for these two waves. This difference becomes critical when these phase shifts approach to π.

To see how the strong SD arises in photonic crystals, let us consider, for example, an electric dipole lattice. Let us try to characterize its electromagnetic response by the effective susceptibility κ, which relates the electric-dipole polarization with the mean field **E** like it is done for continuous dielectric media.

For simplicity let us consider a 2D problem, for example; imagine that these dipoles are dipole moments per unit length of isotropic dielectric cylinders forming a 2D square lattice as shown in Fig. 1.1.

Compare an eigenwave with the wave vectors $\mathbf{q}_1$ and that with the wave vector $\mathbf{q}_2$. Each of these two eigenwaves is a sum of partial plane waves produced by all crystal planes. If $qd < 1$ (a medium without SD or with weak SD) two partial plane waves produced by two adjacent crystal planes are practically in phase, i.e., their interference does not affect κ. Since κ is practically equivalent for the waves $\mathbf{q}_1$ and $\mathbf{q}_2$, this parameter is determined by the near field interactions and does not depend on **q**. Such a lattice does not manifest a spatial dispersion.

At higher frequencies, where $qd \geq 1$ the phase shift of two partial waves produced by two adjacent crystal planes becomes noticeable and the spatial distribution of the eigenwave transforms from a usual plane wave into an interference pattern. It is clear that the interference patters will be different for eigenwaves $\mathbf{q}_1$ and $\mathbf{q}_2$. The ultimate manifestation of this interference is the Bragg phenomenon when the phase shift equals π and the wave cannot propagate. It is clear that the Bragg phenomenon for the wave propagating along the lattice axis and for that propagating diagonally occur at different frequencies. At the Bragg frequency for one wave, the lattice response κ is strong, the effective wavelength λ_{eff} shortens and equals to $2d$, whereas for another wave the lattice response κ is weaker and $\lambda_{\text{eff}} > 2d$. So, κ is different for the waves $\mathbf{q}_1$ and $\mathbf{q}_2$ at the Bragg resonances.

At frequencies slightly below the lowest Bragg resonance the interference of partial waves may also results in the strong spatial dispersion. Namely, if the absolute values of $\mathbf{q}_1$ and $\mathbf{q}_2$ at the same frequency are noticeably different, this means that κ depends on $\mathbf{q}$. Really, in a geometrically isotropic (square) lattice the wave propagates with two different phase velocities in two different directions at the same frequency. Since the refractive index depends on the propagation direction, κ, which is evidently related to the refractive index, also depends on it.

Even if the absolute values of $\mathbf{q}_1$ and $\mathbf{q}_2$ are the same but $qd \geq 1$, the SD is strong. Really, the phase shift between the dipole polarizations of two crystal planes A_1 and B_1 (corresponding to the wave with wave vector $\mathbf{q}_1$) is equal qd. And the phase shift between the polarizations of two tilted crystal planes A_2 and B_2 that corresponds to the wave $\mathbf{q}_2$ is equal to $qd/\sqrt{2}$. Since $qd > 1$ the interference of partial waves in the two cases of the eigenmode propagation is important for κ, and since the phase shifts between two adjacent crystal planes along the propagation direction are different, κ is again dependent on the propagation direction, i.e., on both ω and $\mathbf{q}$. We have to recognize that at high frequencies so that $qd \geq 1$ the lattice is not effectively continuous.

The description of a photonic crystal via the effective susceptibility κ (and effective relative permittivity $\varepsilon = 1 + \kappa$) dependent on $\mathbf{q}$ cannot be called homogenization in the terminology of this book. It is rather an effective-medium model of a photonic crystal. Sometimes this model can be useful if we analytically find the dependence $\varepsilon(\mathbf{q})$. This point will be discussed below.

At lower frequencies ($qd < 1$) one may expand κ and ε into a convergent Taylor's series comprising the powers of $\mathbf{q}$. At these frequencies the interference effects vanish—any eigenmode can be with a high accuracy described as a uniform plane wave. For a plane wave with wave vector $\mathbf{q}$ the series (1.3) can be rewritten as

$$\mathbf{J}(\mathbf{r}) \equiv j\omega\mathbf{P} = j\omega[\kappa_0\mathbf{E}(\mathbf{r}) - j\kappa_1\mathbf{q}\mathbf{E}(\mathbf{r}) - \kappa_2\mathbf{q}\mathbf{q}\mathbf{E}(\mathbf{r}) + \cdots],$$

that is a plane-wave equivalent of (1.3). It implies for $\mathbf{D}$ and ε, respectively:

$$\mathbf{D} \equiv \varepsilon_0\mathbf{E} + \mathbf{P} = \varepsilon_0\varepsilon\mathbf{E}, \quad \varepsilon = 1 + \kappa_0 - j\kappa_1\mathbf{q} - \kappa_2\mathbf{q}\mathbf{q} + \cdots \quad (1.4)$$

The convergence of the series in (1.4) is nothing but effective continuity of the medium. When all terms besides $(1 + \kappa_0)$ are negligible, the SD is absent. Series (1.4) is the alternative definition of SD in any effectively continuous medium. It is adopted in classical books [5, 41, 42, 52, 55, 75, 76], whereas the definition via series (1.3) is adopted in books [38, 40, 53, 54, 56]. It is evident that these definitions refer to the same phenomenon and are equivalent.

Not every term omitted in the right-hand side of (1.4) contains new physics. For known media one may share out the frequency regions for which the series in (1.3) and (1.4) can be truncated on the second order terms. These are frequencies where the medium is effectively continuous. The lowest region of the effective continuity starts at the zero frequency where the only first (zero-order) term in the right-hand side of (1.3) and (1.4) is nonzero. If the constitutive particles have no low-frequency resonances, the lowest region of the effective continuity has the upper bound near the frequency at which $d = \lambda_{\text{eff}}/2$. As a rule, at higher frequencies the medium is also effectively discontinuous and cannot be homogenized. Moreover, the resonances of the constitutive particles may bring one or more frequency intervals where d approaches to $\lambda_{\text{eff}}/2$ to the low-frequency region. At these frequencies the homogenization model may fail. This point will be discussed in details in the last chapter of this book.

Regimes in which the medium is effectively continuous, i.e., the series (1.3) and (1.4) converge but their truncation by the third order of smallness would be not adequate, are practically unknown. An effect of the third order was claimed in [56] for media of strongly helicoidal molecules, but this effect is difficult to detect and it can hardly be resonant. Below we concentrate on the terms of the first and second orders in the effective-medium response. These terms comprise two most important manifestations of WSD: artificial magnetism and bianisotropy.

1.4 Continuity and Locality

1.4.1 Relations by Kramers and Kronig and Their Violation

Hopefully, the reader has already understood that the property called locality is not something absolute. Locality can be perfect or imperfect and the medium can be fully local (no SD), weakly non-local (weak SD) or strongly non-local (strong SD). We have already seen that a *weak non-locality* results in the WSD. In the next chapter we will inspect in details *how* the weak non-locality of an explicit constitutive particle response results in the WSD of a composite material of such particles. Here we will discuss the difference between the concepts of locality and continuity.

Continuity of a medium, once adopted by our model, cannot be partial, weak, or strong. Continuity is either present or absent. It is an absolute property of the medium model. If the model application results for an explicit composite medium are not compatible with the assumption of its continuity, this means that either the model is wrong or the assumption of the continuity was wrong.

How can we judge if the results of our homogenization model are compatible with the initial assumption of the medium continuity? Utilizing the Kramers–Kronig relations to which EMPs of all continuous media must satisfy (see [4, 5, 41, 42, 52–55, 65–67]). These relations hold for all linear continuous media if they are passive (the electromagnetic energy is not generated by their constituents). They follow from the causality principle that for the medium response is linked to the passivity concept. According to H. A. Kramers and R. de Ludwig Kronig, the real and imaginary parts of any component $\varepsilon(\omega)$ of the frequency-dispersive relative permittivity tensor of a continuous medium are related as

$$\text{Re}\,[\varepsilon(\omega)] - \varepsilon_m = \frac{1}{\pi}\int_{-\infty}^{\infty} \frac{\text{Im}\,[\varepsilon(x)]}{x-\omega}\,dx, \tag{1.5}$$

$$\text{Im}\,[\varepsilon(\omega)] = \frac{1}{\pi}\int_{-\infty}^{\infty} \frac{\text{Re}\,[\varepsilon(x)] - \varepsilon_m}{x-\omega}\,dx. \tag{1.6}$$

Here ε_m is the matrix relative permittivity. In the case when $\varepsilon(\omega)$ refers to natural media, $\varepsilon_m = 1$. Analogous relations hold for the effective permeability if the media is a natural or artificial magnetic, and for the parameter of magneto-electric coupling if the medium is bianisotropic. Moreover, all polarizabilities of an optically small particle should satisfy the *microscopic Kramers–Kronig relations* because these relations express the causality of any physical response function in absence of spatial dispersion.

In the corresponding microscopic relations, the left-hand side of (1.5) is substituted by the real part of the microscopic susceptibility and in the right-hand side of (1.5) the imaginary part of the microscopic susceptibility enters instead of Im $[\varepsilon(x)]$. Respectively, the left-hand side of (1.6) will be the imaginary part of the microscopic susceptibility and the right-hand side of this formula contains its real part.

Equations (1.5) and (1.6) were originally derived from the assumption that the bulk polarization **P** of the effective medium is a linear functional of the mean field **E**:

$$\mathbf{P}(\mathbf{r},t) = \int_{-\infty}^{t} \overline{\overline{K}}(t-t')\mathbf{E}(\mathbf{r},t')\,dt', \tag{1.7}$$

where $\overline{\overline{K}}(t)$ is a dyadic susceptibility describing both inertia and dissipation in the electric dipoles of the medium. Causality is expressed by the upper limit t of integration in the right-hand side of (1.7) that was pointed out in the initial study [69] by Kronig. Further, H. A. Kramers in paper [68] used the Fourier transform in the time domain for (1.7) and the standard constitutive relation $\mathbf{P}(\mathbf{r},\omega) = \varepsilon_0(\varepsilon - \varepsilon_m)\mathbf{E}(\mathbf{r},\omega)$. Since the kernel $\overline{\overline{K}}$ in (1.7) is a smooth function of time, the value ε turns out to be an analytical complex function in the upper half-plane of complex frequencies ω. Equations (1.5) and (1.6) were derived by Kramers from the general properties of such complex functions. In cited books one may find the details on the original derivation by Kramers and the generalizations of Eqs. (1.5) and (1.6) to other material parameters—macroscopic and microscopic ones. It was also proved that the Kramers–Kronig relations are not only necessary conditions, they are also sufficient for the causality and passivity of the microscopic response of a particle and for the averaged response of a continuous effective medium.

A seemingly weak point of formula (1.7) is the absence of integration over the volume—it is the same as to assume the kernel $K(\mathbf{r})$ in (1.1) to be proportional delta-function of the distance. Really, the original consideration by Kronig and Kramers neglected the SD. However, if the medium is continuous and its locality is imperfect, we get rid of the integration over the volume transiting from the initial formula (1.1) to formulas (1.2) and (1.3), where the polarization at $\mathbf{r}$ is linked to the field at $\mathbf{r}$ and its derivatives at the same point. The necessity of the volume integration in the calculation of the medium response means only that we have to describe our medium by other EMPs, independent on the dipole susceptibility. For every EMP the response of the continuous medium is purely local, and (if we do the homogenization correctly) all EMPs do not violate the Kramers–Kronig relations.

If these relations are violated for any of these EMPs, and our homogenization model correctly determined all of them as local parameters, we either made mistakes or our assumption of the medium continuity is wrong. In the last case, we have to describe our medium by a more adequate model. If the medium is internally regular, it should be a model of a photonic crystal. For such a medium, we are obliged to find evidence of its electromagnetic discreteness—Fraunhofer diffraction, extra waves inside the slab, or at least a Bragg stopband. If at the frequencies where we have violated the Kramers–Kronig relations such phenomena are absent, and if the isofrequencies in the selected sheer of angles can be modeled as parts of an elliptic or a hyperbolic contour, our medium is effectively continuous. Calculating for this case a set of EMPs violating the Kramers–Kronig relations, we have made mistakes. It is an only possible explanation of the violation of Kramers–Kronig relations for continuous media.

As it was noticed in Preface, dozens of articles published in *Physical Review Letters* in the period 2001–2008 made mistakes in this issue. Taking into account other journals publishing papers on MMs, the number of erroneous papers where the Kramers–Kronig relations were violated is of the order of a few thousands. In the literature on MMs, one can find speculations that the violation of Kramers–Kronig relations is forbidden only for natural materials which are truly continuous, whereas composite media are continuous only effectively. It is evidently a wrong opinion: A continuous medium is a physical idealization in any case, since natural media also consist of particles. There are also speculations that EMPs retrieved from the scattering matrix of the slab using the NRW method (see in Preface) may violate the Kramers–Kronig relations. Really, the medium of the slab can possess a strong spatial dispersion, i.e., can be effectively discrete, whereas the Kramers–Kronig relations are valid only for continuous media. However, it would be better if these authors retracted such papers rather than recognized that all they did was wrong. In fact, the NRW method postulates that the medium is continuous. If the result shows that the medium is discontinuous, the method is not applicable and the whole study is wrong. It is a pity that this simple logic was ignored by these authors and their reviewers.

In the most part of papers reporting the antiresonance in MMs, there are no features of strong SD. From the frequency spectra of the amplitude and phase of both reflection and transmission coefficients of an effective-medium slab, from its refracting properties, from the total internal reflection in thick samples—from all wave processes studied for these MMs, one can understand that they are effectively continuous (at the frequencies of their magnetic resonance band). These papers reported namely a wrong retrieval of EMPs.

Due to numerous publications (especially in *Physical Review Letters*), violation of basic physics in the retrieved EMPs became a commonplace for publications on MMs with artificial magnetism. Some of these works (see [57–59]) claimed that the antiresonance of EMPs was a new physical property of composites with resonant artificial magnetism. However, in the majority of such papers the violation of the Kramers–Kronig relations is simply not discussed.

In 47% of articles on MMs published in 2001–2008 in *Physical Review Letters*, the retrieved EMPs did not manifest the antiresonance. However, it did not obviously mean that the authors knew how to properly characterize these media. In these papers MMs were arrays of optically small silver or gold nanospheres, nanorods, nanopillars, or nanopatches (neither dimers nor oligomers). In such arrays the resonant artificial magnetism is absent. These arrays refer to the class of MMs only due to the so-called *plasmon resonance* of their constituents. A single nanoparticle of Au or Ag in the frequency range of the visible light experiences a resonance of the so-called *localized surface plasmon* mode. This resonance in the optically dense array of such particles may result in the extreme values for the effective permittivity (unusually high or, on the contrary, lower than unity) and in the ultimate anisotropy of the effective permittivity tensor. These features are not observable in the optical range for natural media and makes such arrays be metamaterials. For such MMs a quasi-static homogenization model may be adequate. In some of corresponding papers, the authors even show the agreement between the effective permittivity calculated using classical mixing rules and that retrieved from the ellipsometric or interferometric data. Such MMs simply did not offer to researchers an opportunity to retrieve the antiresonant EMPs and to claim them correct.

Since 2010 the situation in the modern literature concerning the electromagnetic characterization of MMs has improved. However, the antiresonance is not completely defeated even now. Even after relevant publications by mine and by other scientific groups—that of Silveirinha, that of Alú, those of Holloway and Kuester (see the preface)—several authors still insist that the local EMPs may violate the Kramers–Kronig relations. They claim that their EMPs stay beyond these restrictions because result from unusual definitions. In fact, this claim is sometimes (e.g., in works [60, 61]) formally correct. One can introduce EMPs in an exotic way so that at low frequencies they do not transit to their usual static limits. A dynamic homogenization model of a regular lattice can be also developed in an exotic way, and one may define EMPs of a photonic crystal so that they are seemingly local [60]. What is questionable is usefulness of such EMPs. As it was noticed in [55], any useful homogenization model replacing an array of scattering particles by a continuous medium is an approximation. An approximate model can be rigorous only if it allows us to exactly calculate the error in the solution of a boundary problem. This aspect was not analyzed in works [60, 61] and similar papers. Well, in our approximate homogenization model, which stands the limit transitions and delivers a set of EMPs satisfying the Kramers–Kronig relations, the numerical error is also not known a priori. However, our model is, at least, physically adequate and practically work for many finite-size MM samples. As to homogenization models like those suggested in [60, 61], where exotic EMPs are introduced, they are hardly applicable in boundary problems.

This comment does not refer to non-local EMPs of lattices, for which the Kramers–Kronig relations also do not hold. Non-local EMPs can be used in boundary problems (together with additional boundary conditions), and without ABCs are relevant for the analysis of the lattice eigenmodes. Moreover, for non-local EMPs of lattices, the requirements of causality and passivity allow one to establish some restrictions in their dispersion understood as a function of both frequency and wave vector. These restrictions can be called the generalized Kramers–Kronig relations for media with strong SD (see [62]). However, all known effective-medium models of photonic crystals are more difficult mathematically than the theory developed in the present book. Moreover, in the adopted terminology, the effective-medium model of a photonic crystal is not a homogenization model. Therefore, we do not review corresponding works below.

We also do not review the so-called *mesoscopic media*, e.g., *stacked dielectric media* discussed in Preface. It is a group of effectively continuous media with imperfect locality, and the generalized Kramers–Kronig relations for such media have not been derived. Maybe some readers of the present book may be eager to derive these relations for stacked media? Then it would be instructive to learn how it was done for spatially dispersive lattices in work [62].

1.4.2 Lorentzian Dispersion of Material Parameters

As we have understood, if our EMPs are compatible with the idea of a continuous medium they obviously obey the Kramers–Kronig relations. Let our EMPs predict the reflection (R) and/or transmission (T) coefficients for a plane wave of certain polarization impinging the effective-medium slab of the medium under a certain angle correctly, but the Kramers–Kronig relations are violated for them. Thus, our homogenization model is wrong but seems to be useful. Perhaps, this homogenization model is somehow useful and we may ignore the physical adequacy in favor of practice? No, we may not because this approach is practically useless. If the model results in physically meaningless EMPs, it does not correctly describe the electric and/or magnetic responses of the effective medium. What is called in this model permittivity and permeability are fictitious parameters compatible with R and T of a slab only for a specific angle of the plane-wave incidence and given polarization. Such EMPs do not possess a predictive power. They can be successfully applied only to predict the same values of R and T from which they were retrieved. EMPs which are physically meaningless are also practically useless. Our target is to derive a correct set of useful and physically sound EMPs, allowing us to correctly predict R and T and satisfying the Kramers–Kronig relations.

The most common way to satisfy the Kramers–Kronig relations for EMPs of media with WSD is the so-called *Lorentzian frequency dispersion* [74]. Namely, EMPs of all natural continuous media, except plasmas, metals, polar dielectrics, and non-reciprocal and non-linear materials, as well as EMPs of the majority of effectively continuous composites have the frequency dispersion described by the following resonant function:

$$F(\omega) = F_0 + \frac{F_1(\omega)}{1 - \left(\frac{\omega}{\omega_0}\right)^2 + j\frac{\omega}{\omega_0}\nu(\omega)}. \qquad (1.8)$$

Here $F(\omega)$ denotes an EMP, such as effective permittivity or permeability resonating at frequency ω_0. The second term in the right-hand side of (1.8) implies the corresponding medium susceptibility depending on the frequency. F_0, $\nu(\omega)$ and $F_1(\omega)$ are parameters which are either positive constants or, at least, can be approximated as positive constants within the resonance band. Value F_0 represents F in the limit of infinite frequencies. The product $\nu(\omega_0)\omega_0$ is often associated with the so-called *damping frequency* and describes the dissipation in the medium. However, formula (1.8) may also describe the response of an individual scatterer to the local field, e.g., it can be an electric or magnetic polarizability of the particle. In this case $F_0 = 0$ and parameter ν is frequency dependent and describes both dissipation and scattering by the particle. Parameter ν is also frequency dependent in the EMPs of so-called *turbid media*—media in which scattering losses are comparable with dissipation or even higher. In this book, we will concentrate on the effectively homogeneous composites in which the scattering losses are absent. For such media $\nu \approx const(\omega)$ if formula (1.8) describes their EMPs [40, 56, 79, 80, 83].

As to $F_1(\omega)$ when (1.8) describes the effective permittivity of the medium $F_1 \approx const(\omega)$. If (1.8) describes the effective permeability of the medium or a parameter called magnetoelectric coupling responsible for the medium bianisotropy $F_1(\omega)$ must nullify at $\omega = 0$ because these properties disappear in the static limit. Weak frequency dependence means that $F_1(\omega)$ may be approximated by a polynomial, practically, by a quadratic function. Below we will see that our homogenization model (the so-called generalized Maxwell Garnett mixing rule) gives for the effective permeability gives $F_1 \sim \omega^2$ and for the magnetoelectric coupling coefficient $F_1 \sim \omega$.

Only with positive F_1 and ν the expression in the right-hand side of (1.8) satisfies Kramers–Kronig conditions. Antiresonance—shortened *anti-Lorentzian resonance*—implies that F_1 and/or ν are negative in Eq. (1.8). With sign minus for these parameters Eq. (1.8) violates the Kramers–Kronig relations [194].

Now, let us briefly discuss the relationships between microscopic and macroscopic parameters of a bulk array both having the Lorentzian response. In order to describe a bulk array of optically small inclusions/particles in terms of local EMPs the homogenization model has to introduce the material equations (MEs) in a form adequate for this array. These MEs should be valid for any point of inside the array until its effective interface. At the interface they have to transit to MEs of free space

$\mathbf{D} = \varepsilon_0 \mathbf{E}$, $\mathbf{B} = \mu_0 \mathbf{H}$. Then our EMPs allow us to solve the boundary problems, and our model is compatible with Maxwell's boundary conditions. Next, the model has to deduce macroscopic response parameters—medium susceptibilities (second terms in Eqs. (1.8)) from the microscopic response of a particle. To express the EMPs in this way one must know how the local field polarizing the given inclusion and the macroscopic (mean) field are related with one another. We will see in the following chapters that the set of EMPs calculated through the Lorentzian microscopic responses will be also Lorentzian. Vice versa, the microscopic responses properly expressed from the Lorentzian EMPs that were correctly retrieved from R and T coefficients of a medium slab must be also Lorentzian. This retrieval procedure will be also developed in this book.

1.5 About this Book

1.5.1 How Our Theory Is Presented

Since this book is dedicated to the homogenization of media with WSD, its content is basically the homogenization model, and we only briefly mention the wave processes in such media.

After the present introduction, the state-of-the-art is presented in the next chapter. We will review the classical quasi-static homogenization model that in the bulk part is based on the concept of the so-called *Lorentz' sphere* and in the interface part—on the already mentioned extinction principle. To reproduce in this book the original, very long proof of the extinction theorem is not reasonable—it would shift the emphasis from media with WSD to continuous media formed by solely electric dipoles. However, the physical and mathematical content of the theorem is explained with all details needed for understanding.

The further content of this book can be treated as a set of following action points:

- Specify the restrictions of our homogenization model applicability.
- Discuss a few explicit examples.

- Derive relations between the macroscopic medium polarizations and mean fields.
- Derive MEs of media with WSD in the general case via the medium susceptibilities (responses of the unit volume to the mean field).
- Introduce the quasi-static averaging procedure suitable for non-resonant media with WSD.
- Derive equations expressing the medium susceptibilities via the response of an individual particle.
- With an explicit example, study the frequency bounds between the regimes where WSD is negligible and not negligible, and between WSD and strong SD.
- Revisit the classical relations between the local and mean fields and derive the frequency-dependent correction terms.
- Show that for the considered media these MEs are valid up to the effective-medium interface and specify the location of this interface with respect to the particles.
- Consider the special case of an ultimately thin composite layer.
- Study the impact of the resonant artificial magnetism—show that the resonance makes the quasi-static model not applicable.
- Build the homogenization model for resonant lattices of electric and magnetic dipoles.
- Illustrate the self-consistency of this model by explicit examples.
- Suggest the procedure of the EMPs retrieval from R and T of a composite layer.
- Show that the EMPs obtained from the cell problem of an infinite lattice are equivalent to the retrieved EMPs.
- Mimic the violation of Maxwell's boundary conditions by so-called transition layers and retrieve their effective parameters.
- Consider the possibility to squeeze our transition layers to surface sheets responsible for the violation of Maxwell's boundary conditions.
- Consider the extraordinary situation when the medium with WSD is effectively continuous, but our first-principle homogenization model is not self-consistent,

- Modify the first-principle homogenization model for this extraordinary case.
- Write the concluding remarks.

One half of these tasks was performed in [1] based on papers [91–93, 95–97] published before 2003. The second part is based on more recent papers [14–17, 27, 98] and on papers of other scientific teams [21, 22, 24, 25].

1.5.2 Peculiarities of Notations in this Book

When it is possible, notations correspond to the commonly adopted ones. The system of units is SI. Vectors are given by bold font, vector product is denoted as ×, scalar product as ·. Dyads (tensors of the second rank) are denoted by two dashes over the character, triads—by three dashes, etc. Often, we use the index notations for components of the vectors and tensors, where indexes can be Latin or Greek letters. Repeating indexes imply summation over them. For example, the first-order term in the Taylor expansion of the electric field around point **r** presented as

$$\Delta r_k \frac{\partial E_i}{\partial r_k},$$

which means

$$\sum_{k=1}^{3} \Delta r_k \frac{\partial E_i}{\partial r_k},$$

where indices i and k correspond to Cartesian axes. Although the author prefers to denote Cartesian axes by Greek letters, it is not always, helpful because the polarizabilities are also denoted by Greek letters. Hopefully, indices i and j when used for denoting the vector or tensor components will not be mixed by the reader with the imaginary unity and the index k will be not identified with the wave number.

Usually, a planar dyad is presented in the literature as a matrix 2×2, and the usual (bulk) dyad is presented as a matrix 3×3. Respectively, the matrix algebra is used for the operations with the dyads and other tenors. In this book one prefers to present the dyads and other tensors in a string form, that allows the reader not to recall the rules of the matrices multiplication.

An elementary dyad $\bar{\bar{c}}$ in the string representation is a pair of vectors. As a rule, this is the pair of unit vectors of the Cartesian frame (x, y, z) with a scalar coefficient, e.g., $\bar{\bar{c}} = C\mathbf{x}_0\mathbf{y}_0$, where $\mathbf{x}_0$ and $\mathbf{y}_0$ are Cartesian unit vectors. It is a matrix with an only nonzero element—xy-th one—which equals to C. An arbitrary dyad is a linear combination of the elementary ones (four elements in a planar dyad and nine elements in a bulk dyad). It is evident that the scalar product of two elementary dyads $C\mathbf{x}_0\mathbf{y}_0$ and $D\mathbf{y}_0\mathbf{z}_0$ equals $CD\mathbf{x}_0\mathbf{z}_0$. The scalar product of $C\mathbf{x}_0\mathbf{y}_0$ and $D\mathbf{z}_0\mathbf{y}_0$ equals zero. In general, if $\bar{\bar{C}} = \mathbf{ab}$ and $\bar{\bar{D}} = \mathbf{cd}$, we have $\bar{\bar{C}} \cdot \bar{\bar{D}} \equiv \mathbf{a(b \cdot c)d} = G\mathbf{ad}$, where $G = \mathbf{b \cdot c}$. Double scalar product of two dyads $\bar{\bar{C}} = \mathbf{ab}$ and $\bar{\bar{D}} = \mathbf{cd}$ is a scalar $\bar{\bar{C}}\,\mathbf{D} = [\mathbf{a(b \cdot c)}] \cdot \mathbf{d}$. The vector product of two dyads is a triad (tensor of the third rank): $\mathbf{ab} \times \mathbf{cd} = \mathbf{agd}$, where $\mathbf{g} = \mathbf{b} \times \mathbf{c}$. The vector product of $C\mathbf{x}_0\mathbf{y}_0$ and $D\mathbf{z}_0\mathbf{y}_0$ equals $CD\mathbf{x}_0\mathbf{x}_0\mathbf{y}_0$. The vector product of $C\mathbf{x}_0\mathbf{y}_0$ and $D\mathbf{y}_0\mathbf{z}_0$ equals zero.

Scalar product of a dyad $\bar{\bar{C}} = \mathbf{ab}$ by a vector $\mathbf{c}$ is a vector $\bar{\bar{C}} \cdot \mathbf{c} = G\mathbf{a}$, where $G = (\mathbf{b \cdot c})$. Vector product of a dyad $\bar{\bar{C}} = \mathbf{ab}$ by a vector $\mathbf{A}$ is a dyad $\bar{\bar{C}} \times \mathbf{A} = \mathbf{a(b \times A)}$. Hopefully, these examples represent a sufficient illustration of the simplicity of the dyad algebra compared to the matrix algebra.

However, the string representation of tensors is not always optimal. Often, the index form is more relevant, in which an arbitrary dyad is presented as $C_{\alpha\beta}$, where α and β correspond to Cartesian axes. This form covers all 9 (in planar dyads 4) components of the dyad. An elementary dyad has an only nonzero component, e.g., $C_{12} = C$ means $C\mathbf{x}_0\mathbf{y}_0$, because in these notations index 1 corresponds to x, 2—to y, and 3—to z. The index notations evidently generalize to the case of triads, tetrads (tensors of the 4-th rank), etc.

No other special knowledge on tensors of an arbitrary rank except their definition, and the definition of the transposed and inverse tensors are required in order to understand this book. Specific properties of tensors, concerning their eigenvalues, traces, coordinate transforms, Hermitian and non-Hermitian tensor algebra, Jacobians, and Wronskians, are not exploited in this book. The same refers to the group theory of the crystal lattices by Voronoy, Wiegner, and Seitz, to the Brillouin dispersion diagrams, and to all difficult attributes of the solid-state physics such as bandgap structures, periodic potentials, Tamm and Shockley surface states, etc. All questions related to the composite media which are not relevant for the explanation of the weak spatial dispersion are omitted. Formulas in this book, though sometimes involved, are hopefully all quite simple and require from the reader only a basic knowledge of the classical electrodynamics. This simplicity, hopefully, is advantageous for a broad scope of readers.

In this book, everywhere (except some specially commented literature data) the time-harmonic dependence $\exp(j\omega t)$ is adopted.

Special abbreviations used in this book are as follows:

- ABCs: additional boundary conditions
- BA: bianisotropic
- CP: chiral particle
- DC: dichroism
- EMP: effective material parameter
- ME: material equation
- MEC: magnetoelectric coupling
- MM: metamaterial
- m-dipole: magnetic dipole
- OA: optical activity
- OP: Ω-shaped particle (Omega-particle)
- p-dipole: electric dipole
- QME: quasi-material equation
- SD: spatial dispersion
- SRR: split-ring resonator
- TM: transverse magnetic
- TE: transverse electric
- WSD: weak spatial dispersion

Chapter 2

Quasi-Static Averaging of Microscopic Fields and the Concept of Bianisotropy

2.1 View on Weak Spatial Dispersion in the Available Literature

Not in all publications authors clearly distinguish the frequency dispersion and spatial dispersion. We have already understood that these phenomena are very different. Frequency dispersion of the EMP of any medium is not the subject of the present book. Frequency dispersion is inherent to all natural materials and may exist at frequencies where SD is absent—even WSD is not observable. The most common frequency dispersion of natural media is the Lorentzian one that we have reviewed above. Besides it, the Debye dispersion is possible that corresponds to polar dielectrics (see [81]), the Polder dispersion is inherent to ferrites (see [40, 77]), and the Drude dispersion is inherent to plasmas, including the electron plasma in the metals that experiences the Drude dispersion at optical frequencies [71]. Below we study composite media formed by small particles performed of a non-magnetic material. In their frequency dispersion these composites are rather similar to natural non-polar dielectrics, and the frequency

---

Composite Media with Weak Spatial Dispersion
Constantin Simovski
Copyright © 2018 Pan Stanford Publishing Pte. Ltd.
ISBN 978-981-4774-83-3 (Hardcover), 978-1-351-16624-9 (eBook)
www.panstanford.com

dispersion of their constituents is Lorentzian. Their difference from natural media is namely in different manifestations of WSD. For example, artificial magnetism is practically not observed in natural media, and bianisotropy different from chirality is also not observed in them.

In fact, artificial magnetism is, strictly speaking, present in natural chiral media but it is so weak that can hardly be measured. Even the chirality of natural media is not resonant and, therefore, microscopically it is a small effect. However, its order of smallness is first power of the unit cell optical size (qd). The first order of smallness corresponds in the general case to the bianisotropy, and the second order—to artificial magnetism. Terms of the order (qd) in the polarization response are small for natural chiral media but detectable because they result in the rotation of the wave polarization. Terms of the order $(qd)^2$ are negligibly small—there is no way to detect a so small difference of the medium relative permeability from unity. Artificial magnetism is practically important only in composites. There the magnetic response is noticeable and can be even resonant.

Now, let us discuss how the theory of WSD was reflected in the literature prior to the publication of [1]. A quasi-static homogenization model was developed for BA media of electric and magnetic dipoles (see in [79] and [84]). However, this model covers only a special case of BA media—arrays of so-called *canonical chiral particles* and media of so-called *Omega-particles* operating at microwaves. These media will be considered in the present chapter. For other media with WSD the quadrupole response can be significant. In some important cases—especially for nanostructured composites—other high-order multipoles are noticeable in the medium response. In book [1] a quasi-static homogenization model based on the so-called *multipole hierarchy*, applicable to many of these media, was developed.

The idea of multipole hierarchy had been already developed in precedent works of the scientific team guided by R. Raab. Namely, in works [39, 85, 86, 161] the hierarchic multipole expansion was analyzed for molecular media. However, the homogenization model was not built in these works. In these studies the microscopic response of a molecule to the local field was for simplicity identified with the macroscopic response of the effective medium to the mean field. This approximation is adequate only in the optics of gases. It is not adequate even for natural solid media of complex-shape molecule, moreover, it is not applicable to composite media.

It is worth to notice that available publications discussing the manifestations of WSD in media with multipolar response are controversial. Comparing, e.g., works [39, 79, 80, 88] with papers [82, 83, 89] we see different sets of MEs which are physically not equivalent – EMPs adopted in one set of MEs cannot be unambiguously expressed via EMPs of the other set. In some works non-reciprocal BA media are claimed to be media with WSD, and it is claimed that the MEC parameter can be non-reciprocal [79]. In other works it is claimed that the non-reciprocal MEC parameter is prohibited [89].

As to artificial magnetism in media with multipole response, its understanding in different works is especially controversial. Some phenomenological models of artificial magnetism (in works [76, 85, 86]) are compatible with Maxwell's boundary conditions. Some other works such as [87–89] claim that the artificial magnetism does not allow one to apply Maxwell's boundary conditions. In [89] it is claimed that the only manifestation of the second order of WSD is artificial magnetism, and that MEs cannot contain the spatial derivatives of the mean field. In [88] it is claimed that any magnetism has nothing to do with SD, and it is asserted that the SD of the second order yields to the octupole electric response and can be described by the spatial derivatives of the mean field in the set of MEs. Since the homogenization model is not developed in any of these works, all these claim do not look convincing.

This target of this book is to fill in this lacuna presenting a homogenization model that covers, perhaps, not all possible media with WSD but, at least, is physically sound, working for some practically structures and does not contain logical holes and internal contradictions.

It is worth to notice that in the literature on the light-matter interaction there is a physical effect referred to as the effect of WSD which has nothing to do with the content of this book. It is the effect of the so-called *exciton-polariton interaction* in semiconductors. In some meaning, photovoltaic semiconductors, actually, possess WSD resulting in nonzero values of κ_1 and κ_2 in Eq. (1.4). For composite and molecular media these terms origin from the retardation (inertia) in the electromagnetic response of a constitutive particle. The physics of these differential susceptibilities for a photovoltaic semiconductor is totally different.

For semiconductors they result from the combination of two effects. The first one is retardation of the wave over a unit cell of a crystal. The second one is generation of so-called *excitons*—relatively stable electron-hole pairs induced by the light at frequencies laying above the semiconductor crystal bandgap.

Charge excitons correspond to electromagnetic wave packages also called excitons. These wave excitons are movable spots of the quasi-static electric field. They are coupled to *polaritons*—non-propagating eigenmodes of the crystal excited at the interface and attenuating with the depth. The physics of polaritons and excitons is considered in [51, 75, 90] and the theory developed in these works is also called the theory of WSD.

In fact, it is a terminological bug. This theory refers to the solid-state physics and describes the dynamic relations between true (microscopic) fields and charge densities in a semiconductor. The target of the theory is the polarization response taking the excitons and polaritons into account. It is the physics of the solid state and there is no issue of the medium homogenization. The response of a single unit cell of the crystal lattice of a semiconductor has no physical meaning in absence of the lattice, and the concept of the local field is not helpful. Theory of WSD in photovoltaic semiconductors has nothing to do with the theory of WSD in composite and molecular media. It does not allow one to solve a boundary problem without ABCs.

The theory of the present book aims the homogenization model. Our theory of WSD is an explanation how to correctly replace an original composite material or a molecular array by an effectively continuous medium. This is the model in which the weak dependence of the effective permittivity on the wave vector **q** is excluded from the permittivity and transposed into additional EMPs, such as the effective permeability and the MEC parameter. We aim to analytically solve boundary problems for MM slabs without ABCs.

2.2 Introduction to Quasi-Static Homogenization

Consider a bulk array of molecules polarized in the local time-harmonic electromagnetic field. In the classical theory the response of molecules is represented by their electric dipole moments $\mathbf{p}_i$. In the phenomenological explanation, any point $\mathbf{r}$ of the effective medium (modeling the array of molecules) is characterized by the bulk polarization—density of the molecular dipole moments at this point:

$$\mathbf{P}(\mathbf{r}) = \frac{1}{V} \sum_{i=1}^{N} \mathbf{p}_i, \tag{2.1}$$

where N is the amount of molecules in an optically small volume V, centered by point $\mathbf{r}$. In other words, it is assumed that one may arbitrary choose a volume V which is from one side small compared to the wavelength. More strictly, it must be small compared to the characteristic spatial scale of the local field variation. From the other side, it must contain an amount of molecules sufficient for averaging. The mean field is defined as the field averaged so that it practically does not vary across this volume, and varies in space in accordance with the effective-medium eigenmode having an effective wavelength λ_{eff}. In the phenomenological model these words are sufficient – the algorithm of the averaging resulting in such a field is not specified.

The local field acting to an arbitrary selected molecule is postulated to be unambiguously linked to the mean field. Since the mean field does not vary across the volume V, local fields acting to any molecule of this volume are also equivalent and the polarization inside it is uniform (for all i $\mathbf{p}_i = \mathbf{p}$). It results in the equivalence $\mathbf{P}(\mathbf{r}) = n\mathbf{p}$, where n is the molecular concentration at the observation point. The center $\mathbf{r}$ of this volume is characterized by the bulk polarization $\mathbf{P}(\mathbf{r})$ that is proportional to the mean electric field $\mathbf{E}(\mathbf{r})$, and one may write $\mathbf{P}(\mathbf{r}) = \kappa \mathbf{E}(\mathbf{r})$. This local proportionality allows us to treat the array as a continuous effective medium— that with continuous bulk polarization described by susceptibility κ. Further, the phenomenological model does not take care on local fields and microscopic responses any more, and switches to the wave processes in the continuous medium.

In what concerns media with WSD, there is a weak point in the phenomenological approach. Really, this approach implies that the medium unit cell size—inter-particle distance d—is negligible compared to the effective wavelength λ_{eff} in the medium. It must be small compared to the size of the volume V, and the last size should be small compared to λ_{eff}. Therefore, the phenomenological approach would require that d is smaller than λ_{eff} at least by two orders of magnitude. This is respected for natural media even in the optical range of frequencies, and this is why the phenomenological approach was chosen in [5] and similar books. However, even for some natural media the problems with this approach arise due to the multipolar response of molecules [56]. And for composite media with WSD, where λ_{eff}/d is a value of the order of 0.1–0.2, one cannot ignore the transition from the array to the homogenized effective medium. We cannot avoid the consideration of true (microscopic) fields and polarizations. And the first and most important question is proportionality of the local field to the mean field. Let us first inspect, how this proportionality—usually attributed to the names of Rudolf Clausius and Ottaviano Mossotti (see [55, 64, 65])—is proven in the classical works describing molecular media.

2.3 Classical Derivation of the Clausius–Mossotti–Lorenz–Lorentz Formulas

The most known derivation of the Clausius–Mosotti formula belongs to Hendrik-Antoon Lorentz and is based on the use of the aforementioned *Lorentz sphere*. Let us select an arbitrary i-th particle in the bulk of an array of particles. This array can be a regular simple cubic lattice. Also, small deviations from regularity are allowed so that the cubic unit cell $d \times d \times d$ still contains one particle. Let us surround our reference molecule by an imaginary sphere of radius $R_L \gg d$ as shown in Fig. 2.1. Calculating the local field we have to sum up partial fields of all polarized particles $\mathbf{E}_j^{(p)}$ surrounding the i-th one and add this sum to the field $\mathbf{E}^{\text{ext}}$ produced by external sources.

Consider the local field $\mathbf{E}^{\text{loc}}$ at point $\mathbf{R}_i$—that acting on the reference particle assuming that its size a is small enough compared

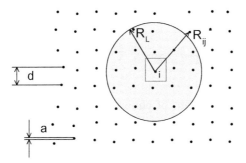

Figure 2.1 Illustration to the concept of Lorentz sphere (of radius R_L). Local electric field E^{loc} at point R_i acts on the particle in the i-th unit cell. The contribution of all particles with $R_{ij} > R_L$ into E^{loc} is replaced by the contribution of the spread polarization $P(R)$. This approach results in the Clausius–Mossotti formula.

to d. Then we can treat the mutual interaction of particles as the interaction of electric dipoles[a]. The local field is the sum of the field produced by the external sources and the partial fields $\mathbf{E}_j^{(p)}$ produced by all (j-th) dipoles but the reference dipole itself at the center $\mathbf{R}_i$ of the reference dipole.

Next, we assume that $d \ll R_L \ll \lambda_{\text{eff}}$. Then the field of the external sources (assumed to be located far from the reference particle) is nearly uniform over the sphere. Therefore, all dipoles induced in the particles located inside the sphere are practically equivalent: $\mathbf{p}_j = \mathbf{p}_i$

The sum of the fields $\mathbf{E}_j^{(p)}$ can be decomposed onto the field of these equivalent dipoles (located inside the Lorentz sphere) and the field $\mathbf{E}_{\text{out}}^{(P)}$ of the bulk polarization $\mathbf{P}$ outside the sphere. Really, the trees of a forest for a distant observer look a continuous greenery. Similarly, the array of molecules located outside the Lorentz sphere is seen by the reference molecule as a continuous polarized medium. Therefore, the contribution of all dipoles located so that $R_{ij} > R_L$ is $\mathbf{E}_{\text{out}}^{(P)}$. This field in the case of the time-varying source varies in space but these variations are noticeable only in the scale larger than the

---

[a]This assumption will be discussed below separately in very details.

sphere radius. Then we have:

$$\mathbf{E}^{\text{loc}} = \mathbf{E}^{\text{ext}} + \sum_{j \ne i}^{N} \mathbf{E}_j^{(p)} + \mathbf{E}_{\text{out}}^{(P)}, \qquad (2.2)$$

where N is the amount of equivalently polarized particles in the Lorentz sphere and the bulk polarization at the surface of the Lorentz sphere is also uniform. The most cumbersome term in (2.2) is the sum of the fields of molecules located inside the sphere.

In his book [74] H. A. Lorentz calculated this sum very simply. Since the amount N of the dipoles inside the sphere is large, the finite sum of the dipole fields can be, according to Lorentz, approximated by an infinite series called in electrodynamics the *dipole sum* [55, 65, 107]. Lorentz has shown that in the case of a simple cubic lattice the dipole sum in the case when all dipoles are equivalent, is exactly equal to zero. Next, he has claimed without a strict proof: if these molecules are randomly displaced from the nodes of the cubic lattice and the regularity is lost but one cubic unit cell still comprises one particle, the result keeps the same [74]. So, Eq. (2.2) simplifies to $\mathbf{E}^{\text{loc}} = \mathbf{E}^{\text{ext}} + \mathbf{E}_{\text{out}}^{(P)}$.

Now let us compare the local field with the mean (macroscopic) field $\mathbf{E}$ at the same point $\mathbf{R}_i$. The mean field is produced by the distant external sources (whose field $\mathbf{E}^{\text{ext}}$ is assumed to be practically uniform across the sphere and does not need to be especially spread) and by the spread polarization of the whole effective-medium sample. The field of the whole homogenized sample at $\mathbf{R}_i$ is the sum of the fields produced by the continuous polarization of the interior of the Lorentz sphere and of its exterior.

Therefore, we have $\mathbf{E} = \mathbf{E}^{\text{ext}} + \mathbf{E}_{\text{in}}^{(P)} + \mathbf{E}_{\text{out}}^{(P)}$ and, for the difference between the local and the mean fields obtain

$$\mathbf{E}^{\text{loc}} - \mathbf{E} = -\mathbf{E}_{\text{in}}^{(P)}. \qquad (2.3)$$

In this difference the fields produced by all the sources located outside the sphere cancel out. The absence of the contribution of the sources located outside the Lorentz sphere into the relation between the local and mean fields is a key prerequisite of quasi-static homogenization. This cancellation means that the macroscopic response at the given point is fully determined by the volume of the Lorentz sphere that is a priori small compared to the wavelength.[b]

In accordance with Eq. (2.3), the difference between the mean and local fields equals to the field produced by a sphere of radius R_L with uniform bulk polarization **P** calculated at this sphere center. The problem of calculating $\mathbf{E}_{in}^{(P)}$ is one of the basic problems of electrostatics, and its solution is well known. The result does not depend on the sphere radius R_L (see [5, 42, 52, 53, 55, 64]) and writes as:

$$\mathbf{E}_{in}^{(P)} = -\frac{\mathbf{P}}{3\varepsilon_0}.$$

Thus, together with H. A. Lorentz, we come to the seminal formula

$$\mathbf{E}^{loc} - \mathbf{E} = \frac{1}{3\varepsilon_0}\mathbf{P}. \tag{2.4}$$

Since the response of a dipole particle is fully determined by the local field at its center and is described by the dyad of the p-dipole polarizability $\overline{\overline{\alpha}}$, relating the dipole moment of i-th particle to $\mathbf{E}^{loc}$:

$$\mathbf{p} = \varepsilon_0 \overline{\overline{\alpha}} \cdot \mathbf{E}^{loc}, \tag{2.5}$$

the bulk polarization **P** of the i-th unit cell is the product of this dipole moment to the concentration of particles in the sample $\mathbf{P} = \mathbf{p}n$. Then, substituting (2.5) and (2.4) to the standard definition of the relative permittivity $\overline{\overline{\varepsilon}}$ of a dielectric medium, namely to the formula:

$$\varepsilon_0 \mathbf{E} + \mathbf{P} = \varepsilon_0 \overline{\overline{\varepsilon}} \cdot \mathbf{E}, \tag{2.6}$$

(in (2.6) the left-hand side is, by definition, the electric displacement **D** [5, 41, 42, 52, 53, 55, 64]) we obtain:

$$\overline{\overline{\varepsilon}} = \overline{\overline{I}} + n\overline{\overline{\alpha}} \cdot (\overline{\overline{I}} - \frac{1}{3}n\overline{\overline{\alpha}})^{-1}. \tag{2.7}$$

Formula (2.7) relates the individual polarizability of a molecule with the effective permittivity of the medium. Here $\overline{\overline{I}}$ is a unit dyad (that

---

[b]This justifies the assumption we have done in the previous chapter when transited from Eq. (1.1) to Eq. (1.2). Our heuristically introduced volume V is nothing but V_L. For the instance, we stay on the Lorentz's point assuming that $R_L \gg d$. However, in the next chapter we will see that if the homogenization is done more strictly V can be identified with the volume of the unit cell.

whose diagonal elements equal unity and other elements are zeros). If the molecular response is isotropic, i.e., polarizability $\bar{\bar{\alpha}}$ is scalar ($\bar{\bar{\alpha}} = \alpha \bar{\bar{I}}$), relation (2.7) takes form

$$\varepsilon = 1 + \frac{n\alpha}{1 - \frac{1}{3}n\alpha}. \qquad (2.8)$$

Eq. (2.8) was phenomenologically predicted prior to H. A. Lorentz by L. V. Lorenz, O. Mossotti and R. Clausius. Notice that Mossotti has suggested its acoustic analogue, and that his original work on this subject was lost. It is known only from private communications. R. Clausius and L. V. Lorenz suggested this formula for the electromagnetic case. However, only work [100] by L. V. Lorenz is available now. The suggestion by Clausius is known from the recollections of H. A. Lorentz, and, apparently, it was the same as that of L. V. Lorenz.

In [100] L. V. Lorenz speculates that the classical polarizability of a sphere filled with a dielectric of permittivity ε and volume V located in free space

$$\alpha = 3V \frac{\varepsilon - 1}{\varepsilon + 2} \qquad (2.9)$$

must be equal to a polarizability of a molecule located in the effective medium with permittivity ε if we substitute $V = 1/n$, where n is the concentration of molecules. Really, a sphere of permittivity ε shared out from a medium of the same permittivity ε being a part of this medium does not scatter. A molecule of a uniform dielectric medium also does not scatter, otherwise such a medium would possess scattering losses and would be turbid. However, we know that it is transparent. Therefore, a medium unit cell with volume $V = 1/n$ centered by a molecule can be modeled by a sphere filled with the permittivity of the effective medium, and its polarizability must be equal to (2.9). Why is it a sphere, and not a cube, if the unit cell is assumed to be cubic? Lorenz explains—a polarized cubic sample as a model of the molecule polarizability does not grant the isotropy to the polarizability, only the sphere grants it.

The weak point of this consideration is the replacement of the discrete medium around the unit cell (replaced by a continuous sphere) by a continuous medium without any convincing explanation why the closely-located molecules act like a continuous medium to the reference molecule. This is why it is called phenomenological model. It deserves to be mentioned because gives the same result as that obtained by H. A. Lorentz in a different way.

In the majority of the literature sources, formula (2.9), equivalent to both (2.4) and (2.8), is called the *Lorenz-Lorentz formula* and formula (2.8) is called the Clausius–Mossotti formula. However, in the literature these two formulas are often mixed up. Therefore, we will call mutually equivalent relations (2.4), (2.8) and (2.9), as well as their analogues for anisotropic lattices (when the cubic lattice depolarization factor $1/3$ is replaced by a tensor value $\bar{\bar{N}}$), the *Clausius–Mossotti–Lorentz–Lorenz* (CMLL) formulas.

Now, let us discuss can we apply CMLL formulas to media with WSD. At a first glance, the answer is negative. First, this model cannot be applied to the vicinity of the medium interface—then the Lorentz sphere will cross the interface. Therefore, we cannot be sure that the interface can be replaced by a continuous medium boundary. Second, the derivation by Lorentz implies that $\lambda_{\text{eff}} \gg R_L \gg d$, and it is not the case of a practical composite medium with bianisotropy and/or artificial magnetism. Third, the derivation of the CMLL formulas by Lorentz is mathematically incorrect. In fact, the zero result for a uniform dipole sum

$$\sum_{j \neq i}^{\infty} \mathbf{E}_j^{(p)} = 0, \quad \mathbf{p}_j \equiv \mathbf{p}_i$$

is wrong. The static uniform dipole sum is equal to zero only for a special order of summation chosen by Lorentz. In fact, this dipole sum depends on the order of summation—one may obtain for it different results from zero to infinity. In other words, the static dipole sum (recall that it is an infinite 3D series) is physically meaningless and must be replaced by a dynamic dipole sum. The dynamic dipole sum converges only if we assume (following to Arnold Sommerfeld) that free space has infinitesimal losses. However, the time-varying dipoles forming an infinite bulk lattice cannot be all equivalent. In accordance with Maxwell's equations, the time variation of the fields and currents in an array results in their mandatory spatial variation. This leads to a nonzero result for a dynamic dipole sum.

If we take into account that the number N of dipoles surrounding the reference dipole within the Lorentz sphere is, in fact, finite and assume that they are all equivalent, the sum of their fields will be also nonzero. In [99] it was shown that this sum even for $N = 100$ is not negligibly small compared to $\mathbf{P}/3\varepsilon_0$ and essentially depends on R_L.

Briefly, the proof of the CMLL formula by H. A. Lorentz was incorrect and the phenomenological speculations were not a proof. However, formulas as such are correct and are applicable not only to natural media but also to many composite materials, including composites with WSD (if only the hierarchy of multipoles is respected for them). For the moment, we simply postulate the CMLL formulas in our theory. Below, an alternative proof of these formulas will be presented. It will be obtained for a bulk of a finite array of particles together with the frequency-dependent correction terms.

In our model, the homogenization of the interface is achieved by extrapolation of the bulk fields and polarizations up to the effective boundary so that Maxwell's boundary conditions are respected. An effective boundary of an array is introduced as a thin layer across which the tangential components of the mean field $\mathbf{E}$ and $\mathbf{H}$ and the normal components of $\mathbf{D}$ and $\mathbf{B}$ will be uniform.

2.4 CMLL Formulas in Optical Theories

2.4.1 Homogenization of Semi-Infinite Crystals

In the literature, the unnecessary restrictions imposed by Lorentz's derivation of the CMLL formulas were discussed earlier and some alternative derivations of these formulas were suggested. However, the most important alternative derivation was that resulting from pioneering works by Erwin Madelung, Claus Oseen and Peter Paul Ewald [101–104]. This way towards the CMLL formulas is important because it was developed for a semi-infinite lattice of dipole scatterers and resulted in the homogenization of the interface.

The general problem was formulated in [101] by Madelung. He was probably the first one, who pointed out that the existing

Fresnel theory of the reflection and transmission for a continuous half-space is not applicable if the continuity is not proved for the interface. The replacement of the original discrete interface of a lattice by a continuous surface is equivalent to the extinction principle—cancellation of the incident wave inside the lattice. If and only if the extinction holds the incident wave interacts with a homogenized lattice as with a continuous half-space. In order to prove the extinction one has to solve the problem of a semi-infinite lattice impinged by a plane wave.

In [101] Madelung considers a simple cubic lattice of isotropic dipole scatterers and an obliquely incident plane wave. He explains that all dipoles referring to any horizontal (parallel to the interface) crystal plane (atomic layer) have the same absolute values because they are all at the same distance from the interface. Only their phases are different if the external wave is incident obliquely. However, in this case the phase varies along the grid from dipole to dipole in a known way. This is so because the incident wave imposes the tangential component $\mathbf{k}_t$ of its wave vector to all waves excited in the lattice. This condition is called the phase synchronism (or phase matching). In continuous media the phase synchronism follows from the continuity of the tangential field vector $\mathbf{E}_t$ across the interface. In periodic structures the phase synchronism is the consequence of the periodicity. It was introduced in the theory of diffraction gratings by Lord Rayleigh [105] and is proved via the decomposition of the total field into so-called *Floquet spatial harmonics*. The phase synchronism is the prerequisite of the unique $\mathbf{k}_t$ for the whole structure that is the key point in the homogenization of the crystal interface.

The complex amplitudes of the dipole polarizations vary somehow in the vertical direction from the top layer of atoms to the bulk, and at a sufficient depth the variation of the dipole moment must be spatially harmonic in order to follow the refracted wave at the wave distances from the surface. The wave vector $\mathbf{q}$ of this refracted wave is unknown and needs to be found. We know only its horizontal component $\mathbf{k}_t$.

Dipole polarizations of all atomic layers can be found using the local field approach which yields the boundary problem to the system of linear algebraic equations. For it let us locate the vertical axis z so that is passes through the centers of some dipoles in every horizontal crystal plane and has the origin at the interface. Call these dipoles located at the arbitrary chosen centers of the crystal planes—points $\mathbf{R}_I = -\mathbf{z}_0 I d$ (here $I = 0, 1, 2, 3\ldots$ and $I = 0$ corresponds to the interface)—the reference dipoles of the I-th crystal plane. The local field acting on any of these dipoles is a sum of the incident wave field $\mathbf{E}^{\text{inc}}(\mathbf{R}_I) = \mathbf{E}_0 \exp(-j\mathbf{k}_0 \cdot \mathbf{R}_I)$ and the partial fields produced at the center of the reference dipole by all dipoles of the semi-infinite lattice but the reference dipole. Partial fields were grouped by Madelung into the fields produced by all crystal planes (J-numbered) including that of the I-th crystal plane. The last one is the so-called incomplete two-dimensional dipole sum. From this 2D series the field of the reference dipole taken at its own center is excluded. Other sums are 2D complete series of the dipole fields produced by regular planar arrays calculated at the center of I-th crystal plane.

The contributions of all crystal planes into the local field $\mathbf{E}_I^{\text{loc}}$ acting on the reference dipole was presented in the form $C_{IJ}\mathbf{p}_J$, where C_{IJ} should have been calculated as explicit functions of on $\mathbf{k}_t$, free-space wave number $k_0 = \omega\sqrt{\varepsilon_0\mu_0}$ (recall that atoms of a crystal are located in free space), lattice period d and the distance $|I - J|d$ between the I-th and J-th crystal planes. The polarizability α of an atom by definition relates the local field $\mathbf{E}_I^{\text{loc}}$ to the dipole moment: $\mathbf{p}_I = \alpha \mathbf{E}_I^{\text{loc}}$. Multiplying this relation by α^{-1} we obtain

$$\alpha^{-1}\mathbf{p}_I = \mathbf{E}_0 \exp(-jIk_0\cos\theta d) + \sum_{j=0}^{\infty} C_{IJ}\mathbf{p}_J, \quad I, J = 0, 1, 2, 3\ldots \tag{2.10}$$

In this equation θ is the incidence angle ($k_{0z} = k_0\cos\theta$) and factors C_{IJ} take into account the same phase distribution of $\mathbf{p}_J^{(mn)}$ in every J-th crystal plane:

$$\mathbf{p}_J^{(mn)} = \mathbf{p}_J e^{-j\mathbf{k}_t \cdot \mathbf{R}_{mn}}, \quad \mathbf{R}_{mn} \equiv (\mathbf{x}_0 m + \mathbf{y}_0 n)d. \tag{2.11}$$

Here the pair of integers (m, n) is the 2D number of a dipole giving the contribution into $C_{IJ}\mathbf{p}_J$, and $\mathbf{p}_J \equiv \mathbf{p}_J^{(00)}$.

Formula (2.11) allows one to express the field produced by the crystal plane through the only dipole moment $\mathbf{p}_J$—that of the atom located at the center of the J-th plane. The factor C_{II} is especially difficult for calculation because the planar dipole sum entering this factor does not contain the term $m = n = 0$.

The approach by Madelung strictly yielded the problem of a semi-infinite lattice illuminated by an obliquely incident plane wave to an infinite system of algebraic equations (2.10). The key point here was the calculation of the planar dipole sum in which the retardation effect should have been taken into account, at least in the first order of smallness, so that to see how important is the retardation for natural crystals. This difficult mathematical problem was only formulated but not resolved by E. Madelung.

A solution was found by P. P. Ewald in work [102] on two conditions: $k_0 d \ll 1$ and $qd \ll 1$ (**q** is the wave vector of the refracted wave). For natural crystals, the condition $qd \ll 1$ is stronger because $q > k_0$. However, the refractive index n of a crystal defined by formula $q = k_0 n$ is not very large. Practically, $n < 10$. Here we may recall that the lattice constant d for the majority of crystals is lower than 0.5 nm and the condition $qd \ll 1$ is obviously satisfied for visible light $\lambda = [400, 800]$ nm.

Ewald's calculation of the discrete spatial distribution of polarizations for all horizontal crystal planes from the top one ($i = 0$) to the deep ones ($i \gg 1$) was done using a semi-analytic approach, based on a specially invented method for summation of the dipole fields, later called the *Ewald–Kummer method*. It allows one to deduce explicit formulas for C_{IJ} in the case $k_0 d \ll 1$ and to find an approximate numerical solution of the system (2.10) in the case $qd \ll 1$. In [103] Ewald shows that the polarizations $\mathbf{p}_I$ of the crystal planes except $\mathbf{p}_0$—that of the interface—follows the refracted wave, i.e., the microscopic polarization in the bulk of a crystal is a spatially harmonic function. As to $\mathbf{p}_0$, it comprises besides a harmonic component also a significant term $\Delta \mathbf{p}$. The significance of this term was explained in [103] and completely clarified in works [102] and [104] where the mean field in the semi-infinite lattice was calculated and compared to the local field. In these works, one has shown (on condition $qd \ll 1$) that $\Delta \mathbf{p}$ is responsible for the cancellation of the incident wave field $\mathbf{E}^{\text{inc}}(\mathbf{R}_I) = \exp(-j\mathbf{k}_0 \cdot \mathbf{R}_I)$ at all points R_I of the lattice starting from $I = 1$. So, the incident wave does not penetrate into the lattice bulk and only an eigenmode of an infinite lattice exists at optically substantial distances from the interface. This eigenmode is the refracted wave whose **q** obeys the Snell's law and whose amplitude and phase is described by the Fresnel formula for the transmission coefficient. Respectively, the reflection really occurs in the top layer of atoms.

The radiation of the top layer of atoms is symmetric with respect to the plane $z = 0$, and the term $\Delta \mathbf{p}$ in $\mathbf{p}_0$ equally contributes into the reflected and transmitted wave field. In the transmitted wave it cancels the incident one. In the reflected wave it creates the field equal to the incident one. Its contribution corresponds to the reflection coefficient equal to (-1). The reflection from a real crystal is not total due to the contribution of the other polarization terms forming the spatially harmonic polarization of the semi-infinite lattice and corresponding to the refracted wave. The sum of the field produced by the peculiar interface polarization $\Delta \mathbf{p}$ and that produced by the spatially harmonic polarization of the whole lattice result in the Fresnel reflection coefficient.

Notice that in both [104] and [102] the extinction theorem was proved not completely strictly. P.P. Ewald *postulated* that $\Delta \mathbf{p}$ is exactly equal to what he needed in order to cancel out the incident wave at points $I = 1, 2, 3 \ldots$. With this assumption he calculated the local and the mean field at the reference points of any crystal plane $I = 1, 2, 3 \ldots$ with the accuracy of the order (qd). The difference of these fields turned out to be equal with the error smaller than (qd) to $\mathbf{p}_I/3\varepsilon_0 d^3$, where dipole moments $\mathbf{p}_I$ were previously found from the system (2.10). This result equivalent to formula (2.4) implied that the CMLL formula followed from the postulated extinction principle. In other words, Ewald has shown the following: if the excessive polarization $\Delta \mathbf{p}$ of the interface layer of atoms really offers the incident wave extinction, CMLL formulas are valid for all atoms, located beneath the interface layer. The validity of these formulas means that the crystal is really a continuous medium starting from its interface.

C. Oseen moved in the opposite direction. Postulating the CMLL formula (2.4) for all atoms in the bulk of the semi-infinite lattice (i.e., for $I = 1, 2, 3 \ldots$), he has analytically solved the system (2.10) and found the excessive polarization $\Delta \mathbf{p}$ of the interface layer. It turned out to be equal to what he needed for extinction. However, in this theory Oseen neglected the terms of the order of (qd) taken into account by Ewald. Therefore, the result by Ewald was more strict.

The collective result by Ewald and Oseen was as follows—the assumption of the incident wave extinction in a dielectric crystal and the assumption that the Lorentz's homogenization model holds for all atoms located beneath the crystal interface are equivalent. Therefore, a semi-infinite dielectric crystal can be really considered as a half-space continuous in the bulk and at the interface. The cancellation of the incident wave inside an optically dense lattice of dipoles is one of the most important theoretical results of the classical optics. All corresponding properties of the continuous medium—the link of the permittivity to the refractive index, the Snell refraction law, and the Fresnel reflection and transmission coefficients—result from the theory by Ewald and Oseen. For finite thickness layers of the medium the extinction principle is applied to both interfaces, and we may homogenize the slab. Thus, the homogenization model matches the electrodynamics of continuous media for finite-size samples.

In both Ewald's and Oseen's proofs the mean field (entering the CMLL formulas) was involved. The strict definition of the mean field was suggested by Ewald in his earlier work [103].[c] It was defined through the so-called *theta-transform* of the true (microscopic) field $\mathbf{E}^{\text{true}}$. This transform allowed Ewald to explicitly get rid of small-scale spatial oscillations of $\mathbf{E}^{\text{true}}$ keeping the oscillations on the spatial scale $2\pi/q$. The theory of the theta-transform makes the original theory rather difficult for readers. Fortunately, later it was proved (in [106] and [107]) that in the case $(qd) \ll 1$ obviously implied in the whole Ewald's theory the theta-transform is with high accuracy equivalent to the simple averaging of $\mathbf{E}^{\text{true}}$ over the volume $V = d^3$ centered at the observation point $\mathbf{R}_i$:

$$\mathbf{E}(\mathbf{R}_i) = \frac{1}{V}\int_V \mathbf{E}^{\text{true}}(\mathbf{R}')dV. \qquad (2.12)$$

The statement that formula (2.12) is equivalent to the Ewald's definition of the mean field can be found in the classical book [54].

---

[c] Ewald was, probably, the first who defined the mean field through the true field strictly.

Although the theory by Ewald was quasi-static—terms of the order of (qd) were omitted in it—the definition (2.12) represents a giant leap towards the dynamic homogenization. Really, the concept of the macroscopic field does not require anymore a *macroscopic* averaging volume (Lorentz sphere) in which the mean field would be uniform. The Ewald homogenization model requires only that the phase shift (qd) per one unit cell is small. The smallness of $(k_0 d)$ in this case follows from the fact that $q > k_0$ for all natural crystals, i.e., the refractive index $n = q/k_0$ of a non-resonant p-dipole lattice is larger than unity. The Ewald model explained why natural media behave as effectively continuous at frequencies of the visible light and at long ultraviolet waves. This was not clear from the Lorentz model.

It is worth to note that in the famous book by Max Born and Eugene Wolf [64] the importance of this result is explained and even a derivation of the extinction theorem is given. However, this derivation differs from the original one. It is incomparably simpler and shorter because the authors replace all the dipole sums $C_{IJ} \mathbf{p}_J$ in (2.10) by the integrals of the continuous polarization. Such a proof is meaningless because this replacement automatically replaces a discrete lattice of dipoles by a continuous medium with a continuous interface. And for continuous interfaces the extinction theorem is not needed—if we postulate that the polarization is continuous, there is no difference between the mean field and true field, we may apply Maxwell's boundary conditions and derive Fresnel reflection and transmission formulas.

In the theory by P. P. Ewald the CMLL formulas were obtained from the extinction postulate. There are alternative derivations of CMLL formulas applicable for bounded arrays of p-dipole and demanding only that d is much smaller than λ_{eff}. These alternative derivations can be found in articles [55, 99, 108, 109] and in book [110]. All these derivations match with the extinction principle. However, no one of these works analyses the bounds of applicability for the CMLL formulas because all these derivations are done within the framework of the electrostatic approximation. The dynamic derivation of CMLL formulas with the analysis of their applicability in different frequency regions will be done in Chapter 4.

The extension of the extinction theorem to non-cubic lattices can be found in works [65, 112]. In [65] the Ewald's consideration was generalized for the orthorhombic lattices when the period d_z of the lattice in the vertical direction z differs from its horizontal period $d_x = d_y$. In [111] the extinction principle is postulated and the CMLL formulas were derived for a semi-infinite lattice with arbitrary anisotropy. In the anisotropic case when the host medium is a dielectric with relative permittivity ε_m the CMLL formula (2.4) generalizes as follows:

$$\mathbf{E}^{loc} - \mathbf{E} = \frac{\overline{\overline{N}} \cdot \mathbf{P}}{\epsilon_0}, \tag{2.13}$$

where the dyad $\overline{\overline{N}}$ is determined by the ratios of the periods d_x/d_z and d_y/d_z. Then we have for the effective permittivity the relation generalizing formula (2.7) to the case when both constitutive particle and lattice are anisotropic:

$$\overline{\overline{\varepsilon}} = \overline{\overline{I}} \varepsilon_m + n\overline{\overline{\alpha}}(\overline{\overline{I}} - n\overline{\overline{\alpha}} \cdot \overline{\overline{N}})^{-1}. \tag{2.14}$$

Tensors N and $\overline{\overline{\alpha}}$ in (2.14) can be found from the solution of two auxiliary problems. The first tensor called *depolarization factor of the unit cell* can be easily found from the analytical or numerical solution of the corresponding static problem since the frequency dispersion for this tensor is not important in the case $(qd) \ll \pi$. The second tensor—anisotropic polarizability of a molecule—can be found in a broad frequency range from the solution of the scattering problem by a single molecule. The tensor $\overline{\overline{\alpha}}$ is the ratio of the induced dipole moment to the time-varying electric field applied to the molecule center. The polarizability is, in general, responsible for the frequency dispersion of the lattice response described by relation (2.14). This dispersion can be noticeable even if the electromagnetic interaction of the lattice molecules is quasi-static. This is so, if these molecules are resonant in the range where $(qd) \ll \pi$ (e.g., artificial molecules).

2.4.2 On the Effect of Surface Polaritons

R. Mahan and J. Obermair in work [113] published half a century after Ewald's and Oseen's works [103, 104] revised their results skipping the consideration of the mean field and CMLL formulas. The local field approach was used in order to solve the system (2.10) in a semi-analytical way taking into account the terms of the order (qd). This analysis of the microscopic polarizations resulted in the important terms which were lost in the theory targeted to the homogenization. Besides small-scale oscillations which are noticeable only on the molecular scale, the polarization of the lattice and the microscopic field comprise some complex spatial harmonics called polaritons. Polaritons were mentioned above in the reference to semiconductor crystals. Here, polaritons arise beyond photovoltaic effects. Polaritons are eigenmodes of the crystal lattice which attenuate versus the distance from the excitation point. In the boundary problem studied in [113] polaritons are excited at the lattice interface by the incident plane wave. Some polaritons called the *staggered modes* have the wave vectors with complex components even in absence of losses [37]. However, in an optically dense p-dipole lattice polaritons in the absence of losses have the imaginary z-component of the wave vector—they exponentially decay versus the distance from the interface and are travelling along it if the wave incidence is oblique [65]. Polaritons decay at a characteristic distance comparable with the lattice period d_z. At the depth of the order of a few d_z (smaller than the wavelength) the polarization of the lattice is spatially harmonic—only the refracted wave survives. This means that the extinction principle holds independently on CMLL formulas.

However, the extinction principle in the formulation by Ewald and Oseen claims: if a medium is continuous in the bulk its interface also must be continuous. If we believe in this claim, the results of [113] for an optically dense lattice of point dipoles allow us to introduce the CMLL formulas. Really, polaritons exist only near the surface and decay at the optically small depth where only the refractive wave survives and the lattice is continuous.

This expectation is confirmed by the study of the reflection coefficient of the lattice. The existence of polaritons in a dense lattice of point p-dipoles is really not harmful for the continuity of its interface—they weakly contribute into the reflection coefficient. They may only have an impact on the reflection phase that can be treated as an effective shift of the reflection plane from the physical interface of the lattice to free space. However, this shift is optically small—of the order of the lattice period—and can be neglected for natural crystals. Further studies have shown that polaritons equally contribute into local and macroscopic fields. This is why they do not destroy the CMLL formulas even for crystal planes $l = 1, 2$ [94]. In any case, the polarization of the interface layer of dipoles noticeably differs from that of the layers located in the bulk (though the polarizabilities of all molecules in this model are the same). This difference is dictated by the surface peculiarity of the local field whereas the mean field remains smooth. This peculiarity is not harmful for the medium homogenization [95], on the contrary, in the model of point dipoles it is its prerequisite.

When [1] was under writing, the author had not known, yet, that the extinction principle holds beyond the case $(qd) \ll \pi$. Now, it is known that it holds even for photonic crystals. It was strictly proved by Pavel Belov and the author of the present book in work [114]. In photonic crystals CMLL formulas do not hold and instead of polaritons and a refracted wave there is a spatial spectrum of propagating eigenmodes. However, even in this case the incident wave cancels out with the wave produced by a few layers of inclusions located on top of a half-space. The incident wave is reproduced by a set of lattice eigenmodes generated in these top layers with the opposite sign. The crucial difference in the extinction principle with the case of an effectively continuous lattice is the *extinction length*—the depth at which the incident wave is completely cancelled. In optically dense lattices the extinction length is the value of the order of d, i.e., is much smaller than λ_{eff}. In photonic crystals the extinction length is the value of the order of several lattice periods d and these periods are of the order λ_{eff}. The extinction length in photonic crystals is optically large and the extinction principle is important only for substantially thick samples.

For the special case of an optically dense dipole lattice the model of [114] is also applicable and degenerates into the Mahan-Obermair model. One may conclude that the cancellation of the incident wave at a certain depth inside a regular bulk array is a general property offered by the internal periodicity. As to slightly irregular bulk arrays, such as liquids the extinction principle also holds. However,

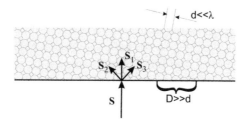

Figure 2.2 A truncated crystal though is optically dense becomes scattering because a new period $D \gg d$ arises on its interface. This new period can be in some cases comparable with the wavelength and attain $\lambda/2$ or λ. If D exceeds λ besides the transmitted ray with Poynting vector S_1 two diffraction rays with Poynting vectors S_2 and S_3 arise. It is clear that homogenization models are not applicable to such a crystal.

up to now it has been proved only for the case when the array is optically dense. This case will be discussed below.

To conclude this subsection we have to mention that the extinction principle was proved only for naturally bounded crystals whose crystal planes are parallel to the interface. Then and only then the period of the structure is equal to d. For truncated crystals—when the crystal planes are tilted to the surface—it is not so. Then the polaritons are coupled with the high-order Bloch harmonics of the lattice eigenmode. These high-order spatial harmonics have negligibly small amplitudes in a non-truncated optically dense lattice. This is why an eigenmode of a usual dielectric crystal is a simple plane wave. In the case of a truncated crystal high-order Bloch modes can be strongly excited because they are coupled to polaritons at the crystal surface. This coupling corresponds to the so-called spatial resonance. This spatial resonance can be interpreted in terms of an effective diffraction grating formed by the surface layer of atoms.

The physical mechanism destroying the homogenization model for a truncated crystal is illustrated by Fig. 2.2. Although in the crystal plane the period d of atoms is optically small, the truncated interface has another periodicity $D \gg d$ determined by the tilt angle. There are special values of the tilt angle for which the corresponding period D becomes comparable with the wavelength—may attain $\lambda/2$ or even be larger than λ. In the first case, we have the

Bragg mode along the surface. In the second case, we have the Fraunhofer diffraction lobes. The scattering waves arise in both reflected field and in the field transmitted into the lattice. For natural crystals this effect is, as a rule, noticeable only for X-rays. An only known exception is micro-crystalline ice. For these small crystals the homogenization model fails in the visible range [115].

2.4.3 On the Impact of Randomness

Only crystal media are truly regular arrays of atoms/molecules. Liquids are slightly irregular lattices of molecules. They can be approximated as regular lattices only in the vicinity of the observation point—on the scale much larger than one unit cell but smaller than the wavelengths of the visible light. Since the lattice of a liquid is not regular on the scale of the wavelength, there is no mutual coherence for the fields produced by distant dipoles at the reference dipole center. This means that the retardation effects are weaker than in a regular crystals for the same unit cell. This allows one to apply the Oseen approach to the problem of a semi-infinite liquid illuminated by a plane wave. In other words, the fields of all p-dipoles in the Madelung model can be replaced by the electrostatic fields. The time-harmonic term $\exp(j\omega t)$ is a common factor and does not influence the result. This theory was developed by D. V. Sivukhin in works [116] and [117].

The model of static dipoles granted a strong simplification compared to the Ewald model in what concerns the proof of the extinction principle. However, the peculiar term $\Delta \mathbf{p}$ in the polarization of molecules located at the interface turned to be more complex. Molecules of a liquid are anisotropic—elongated—and only their thermal movement (including the stochastic rotation) makes an isotropic medium of the liquid. However, the thermal movement of the top molecules in liquids is restricted by the surface tension effect. The reason of the surface tension is the electrostatic interaction of the p-dipoles located on the interface. It orients their dipole moments in parallel—normally to the interface. This effect results in a strong anisotropy in the top grid of molecules. Due to this anisotropy, the vector $\Delta \mathbf{p}$ is oriented not along the local field. This effect implies that the surface polarization obviously has two components—one along the local field and one (the so-called *cross polarization*) is orthogonal. The co-polarized term is responsible for the extinction, and the cross-polarized component (that exists if and only if the wave is incident obliquely) results in the polarization transformation of the reflected wave. Thus, the homogenization model of a liquid in the optical range contains not only the scalar effective permittivity of the bulk but also the surface susceptibility. The last one is a uniaxial dyad, and the difference between its components is more or less noticeable for different liquids in the visible and UV ranges. Especially important impact of this difference corresponds to TM-polarized waves incident with the angle close to the Brewster angle where the cross-polarization of the reflected light is especially noticeable that makes the total transmission (Brewster's effect) impossible [55, 67, 116, 117].

Another example of slightly irregular molecular lattices is the case of some polymer films. There are polymer films (polypropylene and other isotactic amorphous materials) whose molecules are optically small and a surface tension in them is strong. Although their molecules are physically connected, the result of their homogenization is basically the same. Surface susceptibility should be taken into account for the visible and UV light when it is incident under the angle close to the Brewster one [121].

It is worth to note that something similar to Sivukhin's surface polarization was much earlier introduced by Peter Paul Drude for dielectric crystals. Well before Ewald's and Oseen's relevant works, Drude assumed that the boundary problem for a crystal illuminated by a plane wave can be solved with the help of so-called *transition layers*. Following to [118], the transition layer on the crystal surface should be introduced in order to correctly characterize the crystal. Drude pointed out that the Lorentz homogenization model fails near the interface, and the permittivity in this region of the crystal cannot be equal to that of the bulk. However, the Fresnel formulas for the reflection and transmission coefficients hold with rather high accuracy. This means that this region—*transition layer*—must be optically very thin and does not have a strong optical contrast with both free space and crystal. Drude suggested to characterize the transition layer by a certain unknown permittivity and estimated this permittivity based on the evident assumption that it increases across this layer from that of free space (unity) to ε—that of the crystal bulk.

Nowadays, we know that the Lorentz homogenization model works for all crystal planes except the interface, and even at the interface it works for the spatially harmonic polarization term. As to the peculiar term $\Delta\mathbf{p}$, for which this model does not work, it preserves the homogenization for the whole crystal offering the incident wave extinction. Therefore, if we adopt the model replacing the crystal lattice by point p-dipoles we do not need to introduce any transition layers. However, P. P. Drude writing in 1891 his paper [118] and in 1902 his book [67] did not know it. He introduced the local effective permittivity $\varepsilon(z)$ of a transition layer as a scalar varying across the transition layer. The averaged optical response of the transition layer was described by a uniaxial tensor $\bar{\bar{\varepsilon}}_D$ with two different components—tangential ε_D^t and normal ε_D^n. First, Drude assumed that for an orthorhombic lattice of dipole scatterers with periods $d_x = d_y \equiv d_t$ and $d_z = d$ whose interface is parallel to the plane $(x - y)$ the transition layer thickness is equal to d—an only suitable spatial scale in the model of the lattice of point dipoles [67]. Since $(\lambda_\text{eff}/d) \gg 1$ this layer has no practical impact, at least for solid crystals where this ratio has the order 10^3. Second, in a so small interval $\varepsilon(z)$ varying from unity to ε is a linear function. The different components of $\bar{\bar{\varepsilon}}_D$ arise in this phenomenological model because the averaging rules are different for the tangential and normal components of the polarization [67, 118].

So, the impact of the transition layer resulted in some correction term for the Fresnel reflection coefficient. Due to the anisotropy of the transition layer it results in a certain depolarization of the reflected light and modification of the Brewster effect. After the publication of Ewald's and Oseen's works, Drude did not retract his idea. In later editions of his initial monograph [67], Drude insisted that his model is a simple and more accurate alternative to the extinction theorem.

If the Drude model was correct, it should have been also applicable to liquids. The lattice period d for some liquids is much larger than d for natural crystals, and for these liquids the correction term resulting from Drude's ε_D^t non-equal to ε was measurable even in the 1920s. A precise measurement of the reflection coefficient of a liquid could confirm or reject this model. In order to perform this check one needed the normal incidence of the wave, when the cross-component of the surface polarization does not arise (and does not contribute into the reflectance masking the Drude correction term). These experiments were done by the group of Chandrasekhara Raman and reported in works [119] and [120]. Raman's measurements confirmed the higher accuracy of the Fresnel formula. The correction term in the reflection coefficient worsened the accuracy. Therefore, this Drude's model was rejected by the scientific community. In [116] Sivukhin explained this result by a logical mistake of Drude: the thickness of the transition layer d is a microscopic parameter, meanwhile the permittivity is a macroscopic value and cannot vary in the microscopic scale.

In general, this Sivukhin's claim is not convincing. There is no law forbidding a macroscopic parameter to vary over a microscopic interval. Across the sharp interface of the effective medium the permittivity nicely jumps. However, with reference to the Drude model the criticism is justified. In fact, Drude averaged the same lattice response two times in one and the same scale. Indeed, Drude's local permittivity $\varepsilon(z)$ relates the field and polarization averaged over the interval d. Really, the mean field results from the averaging (2.12) of the true field. Next, in order to obtain the averaged tensor $\bar{\bar{\varepsilon}}_D$ Drude performs the averaging of $\varepsilon(z)$ over the same interval d. Definitely, a homogenization model should not average the medium responses twice over the same interval. So, Drude's model in its original variant is not valid. The surface effect in the optical response of solid dielectric crystals is negligible, and the surface effect for liquids results solely from the surface tension and molecular anisotropy.

In the present book, the theory of the Drude transition layer will be revised. We will see that Drude's suggestion makes sense for lattices of finite-size molecules. Indeed, in the approximation of point dipoles there is an only possible model explaining the homogenization of a finite size sample—that based on the extinction theorem. However, if we take into account the finite size a of molecules and inspect the spatial variation on the polarization in this scale we will see how to modify the model by Drude so that to correctly avoid the extinction principle in the homogenization of the interface. In our model, the Drude transition layer is a spread effective boundary of the medium. This spread boundary is raised to free space by $d/2$ from the top layer of molecules. Respectively, a slab of our effective medium comprises an integer amount of unit cells centered by molecules. This model fits Raman's experiments [119, 120] and Mahan-Obermair's theory of surface polaritons [113].

2.4.4 A Bit More about the Anisotropy

Although the classical quasi-static homogenization model is very old and widely discussed in the recent literature readers may reveal some controversy about it. Some authors relate the coefficient $\bar{\bar{N}}$ in the right-hand side of (2.13) (that responsible for the unit cell anisotropy) to the anisotropy of a constitutive particle and equate it to the static depolarization factor of the particle [77, 124, 126, 127]. In the book [125] it is claimed that formula (2.4) holds if and only if a molecule has a nearly spherical envelope. If it was so whatever media of isotropic particles would be isotropic. However, for anisotropic lattices of spherical particles $\bar{\bar{N}}$ does not degenerate into the scalar value $1/3$, it is still a tensor [123]. Vice versa, for a cubic lattice of ellipsoids the tensor $\bar{\bar{N}}$ nicely degenerates into $1/3$ [123]. This misunderstanding definitely origins from the negligence of the homogenization model in popular books. Therefore, the majority of researchers are not familiar with the homogenization model. They know that the phenomenological approach by Lorenz, Clausius and Mossotti works for a cubic lattice of spherical particles and expand it to arbitrary lattices and particles. However, this phenomenology has a weak logical point that we have already discussed. It cannot be extended to non-spherical particles and anisotropic lattices as it is done in [77, 124–127] and many other works.

For whatever anisotropic lattice we have to use the equation (2.14) that for the case of a composite medium with particles in a dielectric matrix of relative permittivity ε_m can be generalized as follows:

$$\bar{\bar{\varepsilon}} = \bar{\bar{I}}\varepsilon_m + n\bar{\bar{\alpha}}(\bar{\bar{I}} - n\bar{\bar{\alpha}} \cdot \bar{\bar{N}})^{-1}. \qquad (2.15)$$

Here tensor $\overline{\overline{N}}$ is determined by the anisotropy of the lattice and has nothing to do with the anisotropy of the particle determined by tensor $\overline{\overline{\alpha}}$. In any case, the anisotropy due to that of the particles and the anisotropy due to that of the lattice can be always taken into account separately. However, up to now these two factors are mixed up by some authors. The results of work [126] turned out to be correct in spite of incorrect arguing and statements because in this paper the spherical scatterers formed a cubic lattice. In this situation both $\overline{\overline{\alpha}}$ and $\overline{\overline{N}}$ degenerate to scalars and the effective medium modeling the lattice is isotropic in the frequency range where it is continuous. Similar comments refer to work [127]. In this work ellipsoids forming an anisotropic orthorhombic lattice are considered. And the Lorentz sphere is generalized into an ellipsoid. The result turns out to be correct because the anisotropy of the constitutive particle (ratio of the ellipsoid axes) and that of the lattice (ratio of the unit cell dimensions) were taken equivalent. In this case the depolarization factor of the ellipsoid is equal to $\overline{\overline{N}}$.

2.5 Maxwell Garnett Model for Dielectric and Magneto-Dielectric Composites

2.5.1 Maxwell Garnett and His Studies of Metal Glasses

Classical works on the homogenization of natural media (by Ewald, Oseen, Sivukhin and others) prove the applicability of CMLL formulas at the optical frequencies for transparent media. For these media the inter-particle distance d is smaller than the effective wavelength at least by two orders of magnitude. Let us repeat two questions we have already formulated above. Are CMLL formulas applicable to composite media—media operating at frequencies where d is smaller than λ_{eff} by one order of magnitude? And if they are applicable, are these media fully continuous—in the bulk and at the surface?

A partial answer based on the experiments was given a century ago. At least for a class of composite media called *metal glasses* these answers were positive. It was done by James Clerk Maxwell Garnett. This scientist was named in the honor of his father's friend, James

Clerk Maxwell and preferred to be called Maxwell rather than James and/or Clerk. He specialized in optics in the 1900s and his doctoral studies were dedicated to the optical properties of metal glasses.[d] Metal glasses had been known since the middle of 19th century. It was a micro-porous dielectric matrix with a colloid solution of Ag, Au or Cr nanoparticles in the pores. In accordance with the modern terminology, a metal glass is a plasmonic MM. In this MM there is no SD and it can be described by an only local EMP—resonant permittivity. The choice of metal glasses as a research subject by Garnett was not random—using a plate of Ag metal glass his father's friend demonstrated in 1855–1857 the first color photograph (this history can be found, for example, in [128]).

Although the arrangement of nanoparticles in a metal glass is random, the uniform distribution of the pores in the glass makes the distribution of nanoparticles practically uniform. The optical contrast between water in the pores and glass matrix is not important, and one may split a sample of the metal glass onto cubic unit cells of volume V, each of them contains one nanoparticle located in the effective dielectric host of permittivity ε_m (permittivity of glass averaged with that of water). The quasi-static homogenization model should be the same as that of a simple cubic lattice. Then why not try the isotropic CMLL formula

$$\varepsilon = \varepsilon_m + \frac{\alpha/V}{1 - \frac{1}{3V}\alpha} \qquad (2.16)$$

and why not substitute the complex refractive index $n = \sqrt{\varepsilon}$ into Fresnel–Airy formulas for the reflection and transmission coefficients of a metal glass plate? Although in the 1900s the theory of the plasmon resonance in metal nanoparticles was not yet built, it was known that the resonance absorption of silver and gold glasses origins from corresponding nanoparticles. Lorentzian dispersion law was already known—the frequency dependence of α could be heuristically modeled as a function (1.8) with $F_0 = 0$.

---

[d]He dropped his research before his PhD was ready and got the second education in international laws. He made a very successful carrier of a barrister and further enthusiastically promoted the idea of the League of Nations. He was elected the first Secretary General of the League Union. Later this Union transformed into modern United Nations.

Using this approach Maxwell Garnett has studied and reported in [129, 130] the reflection and transmission spectra of silver glass layers with different thicknesses and concentrations of nanoparticles, and illuminated under different incidence angles. Although nanoparticles were smaller than the wavelength of the visible light only by one order of magnitude, though these nanoparticles were resonant, and finally, though the CMLL formulas were proved only for the unbounded media, formula (2.16) with the substitution of a Lorentzian function for α matched all experimentally studied dependencies.

Nowadays this result does not seem very surprising for us. First, we know the Ewald homogenization model. This model does not require that the ratio wavelength/unit cell is obviously a value of the order of a few hundred or larger. Second, a lot of other composite media besides metal glasses have been created since the 1900s and the effective electromagnetic continuity of the majority of these composites became a commonplace. However, in the 1900s Maxwell Garnett did not know all this. Moreover, he studied not only substantial metal plates but also composite films with thicknesses as small as dozens of microns. An assumption that a so thin composite film can be described as a layer of a continuous medium for which the CMLL formulas hold everywhere demanded of him a certain courage.

In accordance with Maxwell Garnett, spectral and angular dependencies of the resonant absorption in Ag glass films as well as the dependence of this absorption on the concentration of Ag were very well predicted by the Lorentzian homogenization model. This coincidence was especially surprising because at the resonance the effective wavelength $\lambda_\text{eff} = \lambda/n$ shortens in accordance with the same homogenization model. At the plasmon resonance the ratio d/λ_eff was substantial and attained 0.3 (see also in [63, 122]). How the quasi-static mixing rule (2.16) could keep adequate at the plasmon resonance of a silver glass was unclear, but it was an experimental fact.

The outstanding importance of experiments reported in [129, 130] was commonly recognized. Their main conclusion was applicability of the quasi-static homogenization model to practical finite-thickness electromagnetic composite layers with dipole constitutive inclusions. In honor of the author of these works, the homogenization model represented by the CMLL formulas when applied to whatever composite media is called *Maxwell Garnett model* and CMLL formulas themselves are also sometimes called *Maxwell Garnett mixing rules*.

2.5.2 Maxwell Garnett Model for Magneto-Dielectric Composites

The CMLL formulas can be also formulated for magnetic fields and magnetic dipole moment density. Really, if we adopt (2.4) for an array of p-dipoles we may write its analogue for an array of m-dipoles with moments $\mathbf{m} = \mathbf{M}/n$. Here n is the concentration of m-dipoles and $\mathbf{M}$ is the medium magnetization. The magnetic analogue of (2.4) is, evidently, as follows:

$$\mathbf{B}^{loc} - \mathbf{B} = \frac{1}{3}\mathbf{M}. \qquad (2.17)$$

Although scattering particles with magnetic moment cannot be purely m-dipoles and obviously contain the p-dipole response, Eq. (2.17) is as correct for them as Eq. (2.4). In the quasi-static homogenization model p-dipoles do not contribute into the difference of the magnetic local and mean fields. And vice versa, m-dipoles do not contribute into the difference of the electric local and mean fields. Let us discuss this point involving the concept of the Lorentz sphere.

Let the particles in the Lorentz sphere possess both p-dipole and m-dipole moments. Since we neglect the phase shift between the particles surrounding the center of the Lorentz sphere, the magnetic field of the p-dipoles located at the left from the center exactly cancels out with the magnetic field of the p-dipoles located at the right.[e] Next, the magnetic field of the uniformly electrically polarized sphere calculated at its center is zero. It is so at any nonzero frequency, and results from the problem symmetry. Similarly, the electric field of the m-dipoles located at the left from the center

---

[e] Recall that the magnetic field of an electric dipole is azimuthal with respect to the dipole axis. Therefore, aside of the electric dipole the magnetic field is anti-symmetric.

exactly cancels out with the electric field of the m-dipoles located at the right. And the electric field of the uniformly magnetized sphere calculated at its center is zero at any frequency.

Thus, if the array of particles is a cubic lattice or a slightly random mixture with uniform concentration of particles, in the quasi-static limit p-dipoles do not contribute into the difference $\mathbf{B}^{loc} - \mathbf{B}$. Vice versa, m-dipoles do not contribute into the difference $\mathbf{E}^{loc} - \mathbf{E}$. So, in the homogenization model of p-m particles Eq. (2.4) should be simply complemented by (2.17), and the arrays of the p-dipoles and m-dipoles induced in such particles are homogenized separately. In work [99] an alternative speculation can be found resulting in the same conclusion that also refers to a mixture of p-dipoles and m-dipoles induced in different particles.

These mixing rules allow us to find both effective-medium dyads—that of the permittivity and that of the permeability—if we know the individual electric and magnetic polarizabilities $\bar{\bar{a}}_{ee}$ and $\bar{\bar{a}}_{mm}$, defined by relations

$$\frac{\mathbf{P}}{n} = \mathbf{p} = \varepsilon_0 \bar{\bar{a}}_{ee} \cdot \mathbf{E}^{loc}, \tag{2.18}$$

$$\frac{\mathbf{M}}{n} = \mathbf{m} = \mu_0 \bar{\bar{a}}_{mm} \cdot \mathbf{H}^{loc}. \tag{2.19}$$

Here n is the concentration of particles. Substituting equations (2.18) and (2.19) into (2.4) and (2.17), and taking into account that

$$\mathbf{D} = \varepsilon_0 \mathbf{E} + \mathbf{P} = \varepsilon_0 \bar{\bar{\varepsilon}} \cdot \mathbf{E}, \tag{2.20}$$

$$\mathbf{B} = \mu_0 \mathbf{H} + \mathbf{M} = \mu_0 \bar{\bar{\mu}} \cdot \mathbf{H}, \tag{2.21}$$

we obtain the following set of Maxwell Garnett mixing formulas:

$$\bar{\bar{\varepsilon}} = \bar{\bar{I}} + n\bar{\bar{a}}_{ee} \cdot (\bar{\bar{I}} - \frac{1}{3} n\bar{\bar{a}}_{ee})^{-1}, \tag{2.22}$$

$$\bar{\bar{\mu}} = \bar{\bar{I}} + n\bar{\bar{a}}_{mm} \cdot (\bar{\bar{I}} - \frac{1}{3} n\bar{\bar{a}}_{mm})^{-1}. \tag{2.23}$$

In the scalar version these formulas were derived independently in [55].

We already know that the artificial magnetic particle at microwaves can be a small wire ring that can be made resonant with a split loaded by a capacitance. In the MM literature, this idea is often wrongly referred to work [131]. However, in this paper SRRs were only adapted to the planar technology. With this purpose the wire of a classical split ring was replaced by a printed strip and a dielectric insertion in the gap of the classical ring was replaced by a smaller concentric ring. Such microwave SRRs are definitely more practical if there is no requirement of the isotropic permeability. Classical SRRs were created (see in [47]) for an isotropic mixture in a compound.

2.6 Bianisotropic Media

2.6.1 Introduction to Bianisotropy

Bianisotropic (BA) media are media in which the mean electric field induces the magnetic polarization, and vice versa, the mean magnetic field induces the electric polarization. This phenomenon is also called *magnetoelectric coupling* (MEC).

BA media are characterized by four EMPs that are in the general case dyadic tensors. If BA media are reciprocal three EMPs are sufficient—in this case the tensor describing the electric polarization by the magnetic field and that describing the magnetic polarization by the electric field are basically the same EMP called the MEC parameter. A non-reciprocal isotropic medium with MEC is called *biisotropic medium* and is described by four scalar EMPs. A reciprocal isotropic medium with MEC is called *isotropic chiral medium* and is described by three scalar EMPs.

Biisotropic media with time-harmonic fields are most often described by following MEs:

$$\mathbf{D} = \epsilon_0 \epsilon \mathbf{E} + \sqrt{\mu_0 \epsilon_0}(\kappa - j\chi)\mathbf{H}, \qquad (2.24)$$

$$\mathbf{B} = \mu_0 \mu \mathbf{H} + \sqrt{\mu_0 \epsilon_0}(\kappa + j\chi)\mathbf{E}. \qquad (2.25)$$

MEs of biisotropic media in the form (2.24) and (2.25) were phenomenologically introduced by Ismo Lindell and Ari Sihvola (see [80]) and are, therefore, called the *Lindell-Sihvola* constitutive relations. There are other—earlier—forms of MEs for the same media: the so-called *Post's biisotropic MEs* [41] and so-called *Drude–Born–Fedorov biisotropic MEs* [40]. All these forms are equally suitable for the analysis of the wave processes in such media. The sets of four EMPs entering in these different sets of MEs are different but can be easily expressed through one another. The Lindell-Sihvola formalism is not worse than the two others and is more popular in the modern literature.

Complex EMPs $\xi_D = (\kappa - j\chi)$ in front of **H** in Eq. (2.24) for the vector **D** and $\xi_B = (\kappa + j\chi)$ in front of **E** in Eq. (2.25) for the vector **B** can be called the MEC parameter for the electric displacement vector and the MEC parameter for the magnetic flux vector, respectively. Sometimes, ξ_B is called the *electromagnetic coupling parameter* as a dual counterpart of the MEC parameter. However, this terminology is messy and we will not use it.

In biisotropic media the MEC parameter ξ_D is not equivalent to the MEC parameter ξ_B because $\kappa \ne 0$. Value $\chi \equiv j(\xi_D - \xi_B^T)/2$ is called the *chirality parameter* and value $\kappa \equiv (\xi_D + \xi_B^T)/2$ is called the *Tellegen parameter*. We have already discussed the chirality phenomenon for isotropic materials and know the origin of the chirality parameter. Isotropic chiral media correspond to $\kappa = 0$ and $\chi \ne 0$. Let us now briefly discuss the dual case, when $\kappa \ne 0$ and $\chi = 0$.

In his work [78] Bernard Dominicus Tellegen suggested an idea of a hypothetic composite medium whose constitutive particles possess permanent electric and magnetic moments which are perfectly bounded and parallel to one another. These particles are very light and may easily rotate. In absence of the external field they are oriented randomly and the medium has no polarization and magnetization. In the time-varying electromagnetic field these particles are equally oriented along the electric field vector and along the magnetic one. In other words, the biisotropic non-chiral medium is a polar material that polarizes electrically and magnetically in accordance with (2.24) and (2.25) if we put there $\chi = 0$. Up to now, it is not clear is such a strange medium realizable as an isotropic composite. Most probably, the Tellegen medium is not feasible (see [79]). Similarly, a chiral biisotropic medium with both nonzero χ and κ is hardly feasible.

In the anisotropic case the Lindell-Sihvola equations take form:

$$\mathbf{D} = \epsilon_0 \bar{\bar{\epsilon}} \cdot \mathbf{E} + \sqrt{\mu_0 \epsilon_0}(\bar{\bar{\kappa}} - j\bar{\bar{\chi}}) \cdot \mathbf{H}, \qquad (2.26)$$

$$\mathbf{B} = \bar{\bar{\mu}}_0 \mu \cdot \mathbf{H} + \sqrt{\mu_0 \epsilon_0}(\bar{\bar{\kappa}} + j\bar{\bar{\chi}}^T) \cdot \mathbf{E}. \qquad (2.27)$$

Here and below the index T means the transposed tensor.

Again, tensors of MEC $\bar{\bar{\xi}}_D$ and $\bar{\bar{\xi}}_B$ entering (2.26) and (2.27), respectively, are presented in the form $\bar{\bar{\xi}}_D = \bar{\bar{\kappa}} - j\bar{\bar{\chi}}$ and $\bar{\bar{\xi}}_B = \bar{\bar{\kappa}} + j\bar{\bar{\chi}}^T$ in order to easier distinguish the reciprocal and the non-reciprocal media. Tensor $\bar{\bar{\chi}} = j(\bar{\bar{\xi}}_D - \bar{\bar{\xi}}_B^T)/2$ called the *reciprocal MEC parameter* and tensor $\bar{\bar{\kappa}} = (\bar{\bar{\xi}}_D + \bar{\bar{\xi}}_B^T)/2$ called the *non-reciprocal MEC parameter* or *anisotropic Tellegen parameter*.

Anisotropic Tellegen media, unlike isotropic ones have real chances to be implemented [79]. In work [82] a non-reciprocal BA particle based on a mm-sized ferrite resonator coupled with a metal micro-strip was suggested and there are in [79] some experimental data in favor of this design solution are mentioned. However, in the present book we do not concern such media because we the natural magnetism. This phenomenon is beyond the theory of WSD.

The generalized reciprocity theorem (see [79, 80]) claims that in electromagnetically linear BA structures which do not contain non-reciprocal element (such as ferrites) and elements with optical gain (loads generating the electromagnetic field) the following relations hold:

$$\bar{\bar{\epsilon}} = \bar{\bar{\epsilon}}^T, \quad \bar{\bar{\mu}} = \bar{\bar{\mu}}^T, \quad \bar{\bar{\xi}}_B = -\bar{\bar{\xi}}_D^T.$$

A similar theorem is proved in [40] within the framework of the Drude–Born–Fedorov formalism and in [41] within the framework of the Post formalism. All these proofs are based on the seminal work by Lars Onsager [81].

For reciprocal BA media $\bar{\bar{\kappa}} = 0$ and equations (2.26), (2.27) simplify as:

$$\mathbf{D} = \epsilon_0 \bar{\bar{\epsilon}} \cdot \mathbf{E} - j\sqrt{\mu_0 \epsilon_0}\, \bar{\bar{\chi}} \cdot \mathbf{H}, \tag{2.28}$$

$$\mathbf{B} = \mu_0 \bar{\bar{\mu}} \cdot \mathbf{H} + j\sqrt{\mu_0 \epsilon_0}\, \bar{\bar{\chi}}^T \cdot \mathbf{E}. \tag{2.29}$$

For lossless media all three tensor material parameters $\bar{\bar{\epsilon}}, \bar{\bar{\mu}}$ and $\bar{\bar{\kappa}}$ must be real tensors, as it was proved in [40].

Reciprocal biisotropic media—chiral media were experimentally revealed by Louis Pasteur in 1860 (see more details in [80]). It was already mentioned that chiral media possess optical activity (OA). We have already mentioned that OA means that the phase velocity of the eigenwave with clockwise circular polarization

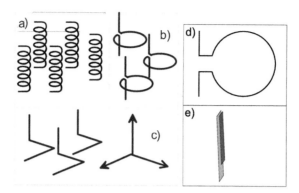

Figure 2.3 Chiral particles—multiturn helices (a), canonical CPs (b), and the Raab hooks (c)—are resonant at frequencies where their overall size is small compared to the wavelength. The Ω-particle (OP) shown in (d) is also resonant at a wavelength much larger than it size. An asymmetric dimer of two parallel metal strips (e) is an analogue of the OP.

differs from the phase velocity of the eigenwave with counter-clockwise polarization. Here we will discuss this property in more details.

The relative difference of the phase velocities in a chiral medium is called the OA parameter. In a chiral medium of right-handed helices the wave with clockwise polarization propagates slower than that with counter-clockwise polarization. Vice versa, in chiral media of left-handed helices the clockwise circular polarization propagates faster than the counter-clockwise polarization.

A linearly polarized wave incident on a chiral medium can be presented as a superposition of two in-phase circular polarizations with opposite rotation. OA implies that the phase shift between these two waves uniformly grows along with the propagation path. The phase shift means the tilt of the resulting linearly polarized vector **E** with respect to the initial polarization plane. Thus, a linearly polarized wave experiences the polarization rotation when transmits through a layer of a chiral medium. A similar turn of **E** holds in ferrites biased by a dc magnetic field. However, as we have already noticed, ferrites are non-reciprocal gyrotropic materials. If a linearly polarized vector **E** (or **H**) after a passage through a nonreciprocal gyrotropic layer in one direction turns clockwise, it turns to the same angle but counter-clockwise when it passes the same path in the opposite direction.

Chiral media are reciprocal and the polarization vector turns in these two cases in the same way. In the medium with right handedness **E** rotates clock-wise for any direction of propagation. It rotates counter-clockwise if the medium is left-handed. For the isotropic chiral medium the polarization rotation does not depend on the propagation direction. In an anisotropic chiral medium the value of the polarization rotation per unit path is dependent on the propagation direction. For example, if all helices are oriented in parallel the polarization rotation is maximal when the propagation direction is orthogonal to their axes and vanishes when the propagation direction is along their axes. However, the reciprocal property of this rotation keeps—it does not change for the opposite propagation direction. The right-handed chirality corresponds to the positive values of χ (more exactly, $\mathrm{Re}(\chi)$, since the medium can have some optical losses making $\mathrm{Im}(\chi)$ nonzero).

We have also mentioned another important property of chiral media called dichroism (DC). DC is the difference in the attenuation coefficients for the clockwise and counter-clockwise eigenmodes. This difference can be very important, when one polarization (clockwise in the right-handed chiral media) attenuates strongly, and the second one (counter-clockwise in these media) attenuates weakly. Very strong attenuation of one polarization occurs if the medium is formed by resonant CPs (in the resonance band also the OA parameter becomes very large). When a wave with circular polarization impinges a substantial layer of a resonant chiral medium (matched to free space, e.g., by an antireflecting coating) it is either fully absorbed in it or almost fully transmitted through it depending on the direction of the polarization rotation in the incident wave. It is clear that such a DC may result in the polarization transformation: a linearly polarized wave after the transmission through the chiral slab becomes circularly polarized because only one circular polarization survives, the second one is absorbed. Unlike the case of so-called *quarter-wavelength plate* and other known polarization transformers this effect holds beyond the interference, does not imply the anisotropy of the medium and holds for a wide sheer of the wave incidence angles. DC is tightly related to

OA because is proportional to the imaginary part of κ, whereas OA is proportional to its real part. However, for resonant chiral media the frequencies at which DC is maximal and OA is maximal can be noticeably different [79].

Recall that both of these effects are united as the phenomenon of gyrotropy, and that for reciprocal media is called chirality. Chirality occurs also in anisotropic chiral media. A BA medium can be referred as an anisotropic chiral medium if its tensor $\overline{\overline{\chi}}$ is diagonal. If this tensor comprises nonzero diagonal components together with off-diagonal components, one tells that the medium is BA and possesses chirality.

2.6.2 Chiral Media, Omega-Media and Their Microwave Realizations

CPs are defined as particles which are anti-symmetric with respect to a certain plane called mirror plane. Here it is worth to note that the term "chirality" origins from Greek $\chi\iota\rho\omega$—palm—because left-hand and right-hand palms are anti-symmetric. Right-handed helices composing a right-handed chiral medium are shown in Fig. 2.3a. Two other variants of chiral particles (CPs) that also may compose a right-handed chiral medium are shown in Figs. 2.3b and c.

A CP shown in Fig. 2.3b is called canonical CP. It was suggested in work [135] and is a parallel connection of a wire ring with a straight dipole scatterer, so that the magnetic moment induced in the loop is parallel to the electric moment induced in the straight wire. A chiral hook depicted in Fig. 2.3b was suggested in work [142]. A particle shown in Fig. 2.3d is called Omega-particle (OP) due to its shape. It was suggested in work [136]. Here the connection of the loop and a wire dipole is in a planar geometry that allows one to easily prepare OPs on a printed circuit board. The bianisotropy of this particle is evident: the electric field along the arms induces the current flowing around the ring and produces the m-dipole. Vice versa, the current induced in the loop by the time-varying magnetic flux flows to the arms and produces the p-dipole. However, in the OP the magnetic dipole is orthogonal to the electric one, and it seems that chiral effects cannot be observed in such media. Arrays of equidistant OPs may compose only BA media without chirality—

tensor $\overline{\overline{\chi}}$ of such media comprises only off-diagonal elements. Media with off-diagonal tensor $\overline{\overline{\chi}}$ are called Omega-media.

BA media with chirality can be formed by mirror-asymmetric oligomers or dimers of OPs [132]. Chirality phenomenon (OA and DC) in these media correspond to the nonzero diagonal components of $\overline{\overline{\chi}}$. As to chirality phenomenon in Omega-media (BA media without chirality) it can be also observed but only for a few selected directions of the wave propagation (see [132]).

To be strict, only in isotropic chiral media the phenomenon of chirality occurs for all directions. In anisotropic chiral media it does not hold for some specific directions. For example, in a medium of parallel helices OA and DC also do not occur if the wave propagates along the axes of the helices. However, in this special case the magnetic field is orthogonal to the helix axis and the helices are not magnetized. In BA chiral media the chirality phenomenon observes for all directions of propagation corresponding to both nonzero polarization and nonzero magnetization of the medium. In Omega-media the chirality phenomenon observes only for selected directions of propagation corresponding to the medium polarization and magnetization.

An isotropic mixture of randomly oriented OPs is a usual magneto-dielectric medium. A so-called *uniaxial Omega-media* formed by dimers of OPs so that $\overline{\overline{\chi}}$ is an off-diagonal tensor has no chirality effect for any propagation direction [79]. It is, anyway, a BA medium and cannot be identified as a magneto-dielectric because its wave impedance depends on the propagation direction. We will recall on this dependence discussing an advanced model of non-BA metamaterials with artificial magnetism.

Here, it is worth to notice that there are complex particles whose properties mimic those of a solid OP. These particles can also compose Omega-media. Such particle is, for example, a dimer of two parallel metal strips of non-equivalent length shown in Fig. 2.3e. The magnetic moment is offered to a dimer by the gap between the strips and the bianisotropy results from the difference of the strips. When the time-varying electric field is directed along the strips it polarizes them differently. The difference of the parallel dipole moments and the nonzero distance between them result in the induced magnetic moment. Also, the difference in the charge distributions along the strips results in their capacitive coupling. The self-inductance of each of two strips and their capacitive coupling result in the resonance for the magnetic mode. The magnetic moment of this dimer is evidently orthogonal to the total p-dipole of the strips.

Notice that an effective medium containing an equal amount of left-handed and right-handed CPs is not chiral. Then the clockwise and counter-clockwise polarizations are equivalent and such media are called *racemic*. Racemic media are magneto-dielectric ones. Artificial magnetism can be resonant in microwave racemic media. Since in natural chiral media the artificial magnetism is negligibly small, the same refers to natural racemic media. They are simple dielectrics [55].

2.6.3 Magnetoelectric Coupling in Metal Bianisotropic Particles

Since we do not consider a colored quartz as an anisotropic chiral medium, we may refer to almost all natural BA media as isotropic chiral media. An only exception are cholesteric liquid crystals— they are anisotropic (uniaxal) chiral media. All natural chiral media exhibit their chirality in the visible range. It is a weak chirality because the MEC parameter χ is not resonant and very low. As to BA composites, the resonance of their MEC parameter—tensor $\bar{\bar{\chi}}$— may be specially engineered and this resonance grants high values of the MEC parameter in the resonance band. To achieve this resonance in an optically small particle one should use a material having the sufficient optical contrast with the dielectric matrix. The highest contrast is offered by a metal. Therefore, BA composites (at least at microwaves) are composites of metal inclusions, as a rule, of those discussed in the previous subsection.

Three suitable geometries of a resonant CP are shown in Fig. 2.3a-c. Particles depicted in this figure are resonant at a wavelength that is much larger than their overall size. Otherwise they could not compose an effectively continuous medium.

Consider a canonical CP impinged by external electric $\mathbf{E}^{loc}$ and magnetic $\mathbf{H}^{loc}$ fields. Vertical component of the local electric field $\mathbf{E}^{loc}_z$ induces in the dipole part of the CP the current which flows into the loop creating the magnetic dipole moment $\mathbf{m} = m\mathbf{z}_0$. In the dipole part the same current creates the z-oriented electric dipole moment $\mathbf{p} = p\mathbf{z}_0$. Vertical component of the local magnetic field H_z^{loc} since it is time-varying induces in the loop the current which flows around the loop creating the z-oriented magnetic dipole moment and in the dipole part, where it creates the z-oriented electric dipole moment. It is, therefore, clear that the m-dipole and p-dipole of the CP are related with the z-components of the local electric and magnetic fields by for scalar parameters:

$$p = a_{ee}E_z^{loc} + a_{em}H_z^{loc}, \qquad m = a_{me}E_z^{loc} + a_{mm}H_z^{loc}. \qquad (2.30)$$

In fact only three of these parameters are independent, because due to the reciprocity (see more details in [137–139]) we have $a_{me} = -a_{em}$.

The loop inductance and the capacitance between the arms of the wire dipole offer the common resonance to both $\mathbf{m}$ and $\mathbf{p}$—both these moments origin from the same current. Polarizabilities $a_{ee,mm}$ attain the maximum of the absolute values at this frequency. Since this resonance is Lorentzian (and since for a polarizability the term F_0 in (1.8) is equal zero), real parts of polarizabilities $a_{ee,mm}$ at the resonance frequency vanish and imaginary parts are maximal. This situation corresponds to the resonant losses in the effective medium formed by such particles. As to a_{em}, it is imaginary for a lossless particle [79], and the imaginary part of a_{em} vanishes at the resonance frequency whereas the real part attains the maximum.

In fact, scalar relations (2.30) represent a rough approximation. This model does not take into account the electric polarization of the loop. A more elaborated model takes into account the non-uniformity of the currents induced in the loop by the applied electric field and by the applied magnetic field [139]. This results in the dyadic relations [137]:

$$\mathbf{p} = \bar{\bar{a}}_{ee} \cdot \mathbf{E}^{loc} + \bar{\bar{a}}_{em} \cdot \mathbf{H}^{loc}, \qquad \mathbf{m} = \bar{\bar{a}}_{me} \cdot \mathbf{E}^{loc} + \bar{\bar{a}}_{mm} \cdot \mathbf{H}^{loc}. \qquad (2.31)$$

Equations (2.31) are general relations for whatever BA particle whose response to the local field can be described as a pair of p- and m-dipoles. In particular, (2.31) hold also for an OP, as well. The main difference from the case of a CP is that a_{ee}^{zz} and a_{mm}^{zz} for OPs are equal zero (i.e., an individual OP does not possess a chiral response). In all cases the reciprocity principle requires that (see [139]) $\bar{\bar{a}}_{me} = -\bar{\bar{a}}_{em}^T$, $\bar{\bar{a}}_{ee} = \bar{\bar{a}}_{ee}^T$, $\bar{\bar{a}}_{mm} = \bar{\bar{a}}_{mm}^T$. Here index T means a transposed dyad.

The fundamental resonance of a scatterer formed by a folded or bent wire occurs when the total length l_{tot} of this wire equals to $\lambda_{\text{eff}}/2$ [79]. This property unites the canonical CP, the OP and the straight wire—if we unfold the wire the magnetic polarization of the scatterer will disappear but the resonance frequency will not change noticeably. To engineer a strong bianisotropy we need to design a constitutive particle with nearly the same absolute values of the p-dipole and the m-dipole. Notice that an m-dipole and a p-dipole can be equated in the normalized variant—if the m-dipole moment is divided by the wave impedance of the space and the p-dipole is multiplied by it. For this equivalence we need a loop of nearly same diameter as the total length of the dipole arms. This is evident intuitively and was confirmed by modeling in [133]. Then the loop radius will be nearly equal to $l_{\text{tot}}/(2\pi + 2) \approx \lambda_0/16$. Here λ_0 is the resonance wavelength in the effective medium. In other words, the overall size of the canonical CP and that of the Omega-particle is nearly eight times smaller than the resonance wavelength. More details on this analytic model and its accuracy for canonical CPs and OPs can be found in book [79].

Now let us discuss the multiturn CP. The resonant properties of a multiturn metal helix were experimentally revealed by Karl Lindman. This scientist in the 1920s decided to emulate OA and DC at ultrashort radio waves scaling the helical molecules of sugar from nanometers to centimeters with a corresponding decrease of the operation frequency [134]. Since a helical molecule is formed by atoms located along the helix the electromagnetic field of the visible light induces in this molecule a discrete polarization current. Moreover, the molecular helix is smaller than the wavelength by two orders of magnitude. Therefore, the chirality of a sugar solution revealed by Pasteur was microscopically very weak.

In the solid metal helix (coil) time-varying electric field applied along the helix axis induces the current which flows around the helix turns and create the m-dipole moment in them. Reciprocally, the time-varying magnetic field creates the current in the turns which flows also along the helix axis creating the p-dipole moment. The resonance of the MEC parameter $a_{me} = -a_{em}$ arises at rather low frequencies because the coil inductance resonates with the inter-turn capacitance. If the pitch angle of the coil is small enough this resonance holds at a frequency at which the individual coil length is very small compared to the wavelength. If the pitch-angle of a helix is rather large (40–50° and more), the resonance holds when the total length of the wire is close to $\lambda_{\text{eff}}/2$. However, even for such pitch-angles the resonant length of a coil several times smaller than λ_{eff} and a sufficiently dense mixture of such coils remains effectively continuous.

It is not surprising that both OA and DC measured for a mixture of wire helices in the experiment [134] reported by Lindman turned out to be much higher than these effects previously measured for the sugar solution in the visible light. The composite by Lindman was the first known chiral metamaterial.

In fact, Lindman's helices as many other complex-shape particles in the resonance range acquire also a quadrupole moment whose resonance also overlaps with those of the electric and magnetic dipole moments. Therefore, their description as well as the general description of any BA particles should be based on the multipole theory, developed in the next chapter of this book. Numerical simulations of multipole polarizabilities of different BA particles operating at microwaves and performed of a thin wire are reviewed in [139]. Resonances of higher multipoles than the electric quadrupole, as a rule, happen at higher frequencies than the p- and m-dipole resonances. High-order resonances in the known cases result in the strong SD and make the effective medium discontinuous. Beyond these resonances the multipole moments of complex-shape particles also may give a noticeable contribution manifested as the WSD.

In the next chapter, we present the general approach to the calculation of multipole polarizabilities for whatever optically small particles taking into account the non-uniformity of the local field. This approach allows us to share out the terms of first and second orders of smallness with respect to the parameter (d/λ_{eff}) in the electromagnetic response of the particle.

2.6.4 Maxwell Garnett Model for a Medium with Both Chirality and Artificial Magnetism

The Maxwell Garnett homogenization model for BA media formed by such scatterers as CPs or OPs whose response yields to the pair of a p-dipole and m-dipole can be easily built by substitution of formulas (2.4) and (2.17) into relations (2.31). Then these relations are substituted into MEs (2.28) and (2.29), taking into account that $\mathbf{p} = \mathbf{P}/n$ and $\mathbf{m} = \mathbf{M}/n$, where n is the concentration of the constitutive scatterers. Details of this algebra for some important special cases can be found in [79].

For example, for an array of parallel canonical CPs with the arms oriented along the z axis, we have $\overline{\overline{A}}_{ee} = a_{ee}^{zz}\mathbf{z}_0\mathbf{z}_0 + a_{ee}^t(\mathbf{x}_0\mathbf{x}_0 + \mathbf{y}_0\mathbf{y}_0)$, $\overline{\overline{A}}_{mm} = a_{mm}\mathbf{z}_0\mathbf{z}_0$ and $\overline{\overline{A}}_{em} = a_{em}\mathbf{z}_0\mathbf{z}_0$. Here the elementary dyad of the transverse electric polarizability $a_{ee}^t(\mathbf{x}_0\mathbf{x}_0 + \mathbf{y}_0\mathbf{y}_0)$ takes into account the electric polarization of the loop (see above). In fact, the electric polarizabilities of a single loop along the x and y axes are different. Therefore for a fully regular array of CPs shown in Fig. 2.3b we should have written $\overline{\overline{A}}_{ee} = a_{ee}^{zz}\mathbf{z}_0\mathbf{z}_0 + a_{ee}^{xx}\mathbf{x}_0\mathbf{x}_0 + a_{ee}^{yy}\mathbf{y}_0\mathbf{y}_0$. However, here we assume some randomness of our array—though all CPs are mutually parallel their loops can be randomly turned around the z axis. Then the loop electric polarizabilities a_{ee}^{xx} and a_{ee}^{yy} average and result in the averaged loop electric polarizability a_{ee}^t. After mentioned substitutions we obtain:

$$\overline{\overline{\varepsilon}} = \left(\varepsilon_m\varepsilon_0 + \frac{na_{ee}^t}{1 - na_{ee}^t/3\varepsilon_m\varepsilon_0}\right)\overline{\overline{I}}_t$$

$$+\mathbf{z}_0\mathbf{z}_0\left[\varepsilon_m\varepsilon_0 + \frac{1}{\Delta}\left(na_{ee}^{zz} - \frac{n^2}{3\mu_0}(a_{ee}^{zz}a_{mm} + a_{me}^2)\right)\right], \quad (2.32)$$

$$\overline{\overline{\mu}} = \mu_0\overline{\overline{I}}_t + \mathbf{z}_0\mathbf{z}_0\left[\mu_0 + \frac{1}{\Delta}\left(na_{mm} - \frac{n^2}{3\varepsilon_m\varepsilon_0}(a_{ee}^{zz}a_{mm} + a_{me}^2)\right)\right], \quad (2.33)$$

$$\overline{\overline{\kappa}} = -\frac{\mathbf{z}_0\mathbf{z}_0}{\Delta}na_{me}, \quad (2.34)$$

where it is denoted:

$$\Delta = \left(1 - \frac{na_{ee}^{zz}}{3\varepsilon_m\varepsilon_0}\right)\left(1 - \frac{na_{mm}}{3\mu_0}\right) + \frac{n^2 a_{me}^2}{9\varepsilon_m\varepsilon_0\mu_0}.$$

In these formulas ε_m denotes the permittivity of the dielectric matrix into which CPs are embedded. Individual polarizabilities of a CP are denoted as $\bar{\bar{a}}_{ee,mm,em,me}$ and $\bar{\bar{I}}_t = \mathbf{x}_0\mathbf{x}_0 + \mathbf{y}_0\mathbf{y}_0$ is a unit planar dyad in the transverse (to the z-axis) plane ($x-y$). Formulas (2.32), (2.33), and (2.34) will be revised below and replaced by more general formulas.

Expressions (2.32), (2.33), and (2.34) keep the Lorentzian dispersion in EMPs if it is inherent to the individual polarizabilities of a CP. It is very difficult to prove analytically with these equations, but readers may check that it is so in a numerical exercise, introducing reasonable parameters of the Lorentzian dispersion for the individual polarizabilities and reasonable concentration of CPs.

Below, in Chapter 3 an analytical proof will be presented when the Maxwell Garnett mixing rules are generalized for media with multipole response.

The overall size of a typical BA particle operating at microwaves is close to $a = \lambda_0/8$ and the inter-particle distance d is evidently larger than a. Therefore, in the resonance band the unit cell of the effective medium is in all practical cases as substantial as one quarter of the effective wavelength. Is such an effective medium sufficiently optically dense? Is the homogenization model for such a composite valid for the resonance band? The metal glass in the experiments by Maxwell Garnett was effectively continuous, but the constitutive particles were as simple as spheres, and the effective medium was a simple dielectric. BA media of CPs or OPs possess both artificial magnetism and bianisotropy, and these properties are resonant. Are we sure that such media are still effectively continuous?

There is no general answer, but as a rule it is so for practical microwave metamaterials. Full-wave simulations and experiments (see [83, 139–141]) have shown that the practical composites of canonical CPs and OPs behave as homogeneous media in the whole resonance band. Notice that this band is rather broad (relative bandwidth is 5–10%) because the effective medium of microwave CPs and OPs is, as a rule, characterized by rather significant losses even if the dielectric host is practically lossless and the BA particles are fabricated of a substantially thick copper wire. Losses in the particle are so high at the resonance because the electric field locally concentrates in the gap between the arms and the dissipation in the metal magnifies dramatically.

2.7 Some Restrictions of Our Study Subject

2.7.1 Why We Do Not Consider the Condensed Composites of Complex-Shape Metal Particles

In this book we do not consider arrays of densely packed metallic particles. However, it is worth to mention that the most popular model of quasi-static homogenization for densely packed arrays of simple-shaped particles was suggested by Dirk-Anton von Bruggeman [143]. The Maxwell Garnett model is adequate for rather dilute composites—$d > 1.5a$, which usually corresponds to the volume fraction of particles in the matrix lower than 0.3. The Bruggeman model is developed for the case when this fraction is as high as $0.3\ldots0.6$. This model treats the gaps between the densely arranged inclusions (those filled with the host material) as particles of the second type. These mutually touching dielectric particles compose the same effective medium in the matrix formed by the inclusions (metal ones or performed of another dielectric). Using such speculations one may obtain an alternative set of mixing rules for the isotropic and anisotropic cases. Initially, this approach was developed in the 2D variant for a densely packed array of parallel cylinders [143]. In work [84] this theory was extended to the arrays of parallel ellipsoids, and in [43]—to densely packed parallel helices. In [146] one successfully applied the Bruggeman approach to cholesteric liquid crystals operating in the resonance band. However, in the majority of works dedicated to resonant composites researchers prefer to use Maxwell Garnett mixing rules even for densely arranged composites, claiming that the Bruggeman approach is not adequate for complex-shaped particles and its applicability bound is restricted by frequencies below the fundamental resonance of the constituents (see [79, 83, 144]).

Besides the Bruggeman model, for densely packed arrays one developed other quasi-static models called *percolation models*. They are applicable to the case, when particles (usually metal ones) touch

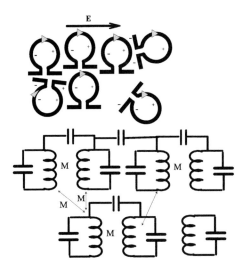

Figure 2.4 An in-plane group of densely packed OPs (top) and its equivalent scheme (bottom). A densely packed effective medium can be composed by a set of effective sheets formed by such clusters.

one another, forming effective chains and even effective stochastic networks (see [63, 147–149]). Of course, static homogenization models for simple-shape densely packed optically small inclusions are adequate if and only if the retardation effects are negligible for them.

In densely packed media with WSD, e.g., in materials with artificial magnetic properties, the retardation effects evidently cannot be negligible. Such media, of course, can be also densely packed and complex-shape particles such as SRRs may touch one another or even spatially overlap. For these media the Bruggeman model is not suitable but for us such materials are not interesting and we will not try to homogenize them. To understand why these media are not interesting for us let us inspect an array of densely-packed Omega-particles.

The sketch in Fig. 2.4a shows a cluster of OPs shared out from a densely packed effective Omega-medium. In this cluster there are no separate particles, no scattering centers. The polarizabilities of an individual particle are not relevant. Mutual capacitance effectively connecting adjacent particles in the equivalent scheme Fig. 2.4b can be even higher than the capacitance between the collinear arms of an individual OP. Mutual inductances M between adjacent OPs laying in the same plane are lower than the self-inductance but there are also OPs located in the parallel planes and when their loops overlap their mutual inductances may be comparable to the self-inductance. Moreover, occasional touching of adjacent OPs may result in the shortcuts of their arms and loops. All these connections cannot be described analytically.

In principle, such a medium can be homogenization numerically without local fields, and such a homogenization may be correct and physically sound. The spread bulk densities of electric and magnetic currents can be introduced and related to the mean electric and magnetic fields by some coefficients. These coefficients can be found for a given array after a numerical simulation of a certain sufficiently large cluster, whose interior contains typical features of the whole array. One may try to fit these simulations to the heuristic model.

Attempts of such a numerical homogenization are known (see book [63]). Moreover, below we discuss namely this homogenization model in order to better outline the advantages of the multipole descriptions. Here, it is enough to point out that such homogenization models may be important only for the designer of absorbers. These composites are not transparent. A large volume fraction of metal wires and a strong concentration of microscopic electric fields in the numerous capacitive gaps implies huge conductive losses. Moreover, if the internal geometry is random, as that shown in Fig. 2.4a these conductive losses are complemented by high scattering losses. The refracted waves in such media will decay very fast. Such media are similar to those based on classical SRRs by Schelkunoff and Friis [149]. In the present book we prefer to concentrate on media which are sufficiently transparent so that some wave processes different from the wave dissipation may hold in them. For it we need the condition $(d/a) > 1$, i.e., all metal particles must be spatially separated so that one may split the composite onto unit cells with one particle per cell.

Of course if the adjacent particles are too distant compared to their sizes, i.e., $d \gg a$ they modify the material parameters of the host medium weakly. It is evident that there is a certain optimum of concentration of particles for which the high transparency of the medium is achieved simultaneously with the high level of the artificial magnetism and/or bianisotropy. For OPs and CPs operating in the microwave range, this optimum corresponds to the mean inter-particle distance d which is nearly twice as large as the overall particle size a [150, 151].

2.7.2 On Composites of Dielectric Particles

Not only arrays of elements performed of curved metal wires, metal strip dimers, and other metal scatterers can possess the resonant artificial magnetism. Arrays of simple-shape dielectric inclusions may also manifest this phenomenon in both radio-frequency and optical ranges.

Usually, dielectric inclusions for effectively continuous composites with artificial magnetism are spheres since a dielectric sphere has a stronger scattering resonance than an ellipsoid or a particle of another shape. Respectively, arrays of spheres performed of ceramics (at microwaves) and of semiconductor materials (in the optical range) can be resonant due to the scattering resonances of an individual sphere. Scattering resonances correspond to the resonant p-dipole, resonant m-dipole and higher resonant multipoles induced by the external wave field. These resonances are called *Mie resonances* upon Gustav Mie who, for the first time, strictly solved the problem of the plane-wave diffraction by a sphere. Mie has represented the field scattered by a dielectric sphere as a series of multipole fields (see more details in [64]) that allowed the multipole resonances to be studied separately. The lowest resonance frequency of an optically small high-permittivity sphere is that of the m-dipole. The next Mie resonances are those of the p-dipole, magnetic quadrupole, electric quadrupole, etc.

The condition of the magnetic Mie resonance is $a \approx \lambda/n_s$, where n_s is the refractive index of a sphere (referred to that of the host medium) and λ is the wavelength in it [64]. It is clear that the sphere at this resonance can be optically small even with respect to the effective wavelength in the composite medium if $n_s \gg 1$. This makes the magnetic Mie resonance attractive for creating the artificial magnetism. The internal electric field at this resonance is distributed so that its lines form a set of circular loops, whereas the magnetic field is concentrated at the axis of the sphere parallel to the external magnetic field. In the band of the magnetic Mie resonance the effective permeability μ of a optically dense array can be noticeably different from unity and, in accordance with the Lorentzian dispersion law (1.8), can even become negative at some frequencies. Therefore, in the band of the resonant artificial magnetism the effective medium may be opaque [152].

The transparency in the range where $\text{Re}(\mu) > 0$ is determined by the complex refractive index $n = \sqrt{\varepsilon\mu}$, which has the imaginary part strongly varying within the magnetic resonance band. In fact, the magnetic resonance band of such a composite can be split onto frequency intervals. Within one intervals the medium is transparent, within the other intervals it is opaque. In spite of this complexity at these frequencies such an array can be adequately described as a composite formed by particles having only the p-dipole and m-dipole responses. The p-dipole is not yet resonant at the magnetic Mie resonance but is noticeable even beyond the resonance and its contribution is important. The non-resonant electric polarization of the spheres can be described by their permittivity $\varepsilon_s = n_s^2$. Therefore, the permittivity of the spheres can be mixed with the host medium permittivity using the heuristic mixing rule: $\varepsilon_{\text{eff}} = f\varepsilon_s + (1-f)\varepsilon_m$, where f is the volume fraction of spheres. After this modification of the host medium, we may consider the composite structure as an array of m-dipoles in a uniform host. This approach will be used in the last chapter of this book.

At the next Mie resonance of a sphere—that of the p-dipole—the magnetic moment becomes negligible. At this resonance the medium can be still effectively continuous, if n_s is sufficiently high and the spheres are arranged with an optically small period. At frequencies below the magnetic Mie resonance and in between the bands of the magnetic and electric ones the Maxwell Garnett model can be adequate [153]. This is so if $n_s \gg 1$, practically if $n_s > 3$–4.

Examples of an effectively continuous array of dielectric spheres for which the Maxwell Garnett model or Bruggeman model would be applicable at the electric Mie resonance are also known. However, as a rule the Maxwell Garnett homogenization (and moreover the Bruggeman homogenization) is not adequate at the magnetic resonance frequencies. Although in the well-known work by L.

Lewin [152] the Maxwell Garnett model was applied for such a composite within its magnetic resonance band, this model was not confirmed by full-wave simulations or measurements. Although this paper was seminal for the development of MMs, the Maxwell Garnett homogenization model is not adequate for the array suggested in [152]—it does not stand the validation by full-wave numerical simulations [154, 155].

Quasi-static homogenization is applicable for such MMs only in some special cases. This is, for example, the case of a cubic lattice of polaritonic (lithium tantalate) spheres operating in the mid infrared range [156]. This is also the case when the cubic lattice is composed by spheres of ferroelectric ceramics (so-called *BST material*) operating at microwaves [157]. However, in both these cases (when the quasi-static model works) optical losses are so high in the whole resonance band that the medium can operate only as an absorber.

A lattice of densely-packed spheres performed of a hypothetic magneto-electric material was theoretically considered in [158]. Here the Maxwell Garnett homogenization model matches the full-wave simulations in one half of the resonance band. However, this partial adequacy of the quasi-static model can be attributed to exotic values of the material parameters of spheres. Isotropic magnetic materials with so high permeability and low losses as adopted in [158] do not exist in nature. As to the feasibility of such spheres as samples of a composite medium, it is also doubtful. In [158] the permeability of spheres is assumed to be dispersion-free that is hardly possible for a passive composite with so high permeability.

In work [159] the authors claim that the quasi-static homogenization model fits the experiment and full-wave simulations for a low-loss MM performed of realistic high-permittivity spheres. This MM is characterized by both strong artificial magnetism and high transparency. This extraordinary result, however, can be explained simply. The formulas describing the quasi-static homogenization in [159] are only seemingly quasi-static mixing rules. They are neither the Maxwell Garnett formulas nor the Bruggeman formulas (the author of the present book has checked that both Maxwell Garnett model and Bruggeman model give very different predictions). These formulas were introduced in [159] heuristically and only *look* like a valid quasi-static homogenization model. In fact, this is not a homogenization model at all, but a fitting procedure that relates the data strictly obtained for the medium macroscopic response to the parameters of the original spheres. The physical meaning of the introduced coefficients remains unclear.

So, based upon the analysis of the available literature, one may conclude: for effectively continuous arrays of high-permittivity spheres the quasi-static homogenization models within the band of the magnetic Mie resonance are suitable only in the case when the effective medium is sufficiently lossy and cannot be called transparent. For low-loss composite media of highly-refractive spheres the quasi-static homogenization models in the band of the Mie resonance are not applicable. Such media are MMs with resonant artificial magnetism.

Chapter 3

Multipolar Theory of Weak Spatial Dispersion

3.1 Preliminary Speculations

Mean (macroscopic) electromagnetic field is a pair of vectors (**E**, **H**) entering the MEs of any effective medium and—through these MEs—the macroscopic Maxwell's equations written for the medium. Mean fields result from averaging the true (microscopic) fields ($\mathbf{E}^{\text{true}}$ and $\mathbf{H}^{\text{true}}$) created by external sources and by all components of the medium—constitutive particles and matrix (if different from free space). In phenomenological models, no attention is paid to the procedure of averaging and the concept of the mean field is intuitive. Readers of the present book have to understand the bounds of validity of the homogenization model at least to the end of this book. Therefore, a clear definition of the procedure of averaging the true fields and polarizations is necessary.

Such a relationship was introduced by Ewald for a regular lattice of p-dipoles. Unfortunately, in that model the mean field and polarization were calculated only at the lattice nodes $\mathbf{R}_i$. Is it not enough for our purposes. Our homogenization model—multipolar theory—needs to consider the fields also beyond the lattice nodes.

Composite Media with Weak Spatial Dispersion
Constantin Simovski
Copyright © 2018 Pan Stanford Publishing Pte. Ltd.
ISBN 978-981-4774-83-3 (Hardcover), 978-1-351-16624-9 (eBook)
www.panstanford.com

One of our targets is to obtain MEs valid at every point of the space and compatible with Maxwell's boundary conditions. Moreover, our theory should be applicable to both regular arrays and slightly irregular ones.

Consider the difficulty related with the extrapolation of the homogenization model to the points beyond the lattice nodes on the example of a dipole lattice. The bulk dipole polarization $\mathbf{P}$ is expressed in this model through the dipole moments of particles for the observation point located at the lattice node simply: $\mathbf{P}(\mathbf{R}_i) = \mathbf{p}(\mathbf{R}_i)/V = n\mathbf{p}(\mathbf{R}_i)$, where V is the volume of the unit cell and n the concentration of particles.

However, which dipole moment do we have to substitute if $\mathbf{R} \neq \mathbf{R}_i$? That of the closest particle? Or do we have to average the dipole moments of all adjacent particles surrounding the observation point? Of course, if the inter-particle distance d and, therefore, the phase shift (qd) of the wave propagating in the effective medium are negligibly small, both these approach result in the same $\mathbf{P}(\mathbf{R})$.

If the value (qd), though small, is not negligible, e.g., when $(d/\lambda_{\text{eff}}) \sim 0.1$–$0.2$ the question becomes actual and one of two possible approaches should be selected. Here, we have to reject the idea of identifying $\mathbf{p}(\mathbf{R})$ with the averaged dipole moment obtained through dipoles surrounding the observation point $\mathbf{R}$. In this case the volume of averaging V will be too large and such a homogenization model will be irrelevant. In fact, averaging should spread only subwavelength oscillations of the field and polarization but must keep the spatial oscillation characteristic for the eigenwave.[a] Therefore, the volume of averaging centered by the observation point should be equal to that of the unit cell. In order to avoid the ambiguity we may define the bulk polarization $\mathbf{P}(\mathbf{r})$, first, at the points $\mathbf{R}_i$—centers of the array particles—and then consider the function $\mathbf{P}(\mathbf{r})$ as a result of a smooth extrapolation to the whole space. The explicit type of this extrapolation (linear, cubic, spline) is not important for our model. The same can be thought about all multipoles induced in the particles of the array.

---

[a]If we choose the averaging volume V so that its size equals to the effective wavelength λ_{eff}, the electromagnetic field will nullify after such an averaging. Identically zero mean fields will be hardly helpful for solving the electrodynamic problems.

Our homogenization model is not obliged to be rigorous, but it must be physically meaningful and compatible, if possible, with Maxwell's boundary conditions. It would be great if it were also rigorous and allowed us to calculate the error in the solution of boundary problems. However, this is not a realistic target. A useful for an approximate solution and physically adequate homogenization model of bulk arrays of realistic scatterers is a realistic task.

Next, let us discuss the target of our homogenization model—a set of local EMPs. These EMPs do not reveal too detailed information on the structure geometry. Of course, such general features as the anisotropy and/or bianisotropy should be clear from the EMPs. However, the shapes of particles and the details of the lattice geometry cannot be revealed from local EMPs. For example, a simple cubic lattice and a face-centered cubic lattice of dipoles should both be described by the scalar permittivity and from its magnitude and frequency dependence the type of the cubic lattice cannot be seen. The target of the homogenization is to avoid in the numerical simulations and in the analytical calculations an amount of excessive information on the material internal structure. Our homogenization is local, i.e., we target our homogenization model to be applicable without additional boundary conditions.

3.2 Main and Auxiliary Vectors of the Macroscopic Electromagnetic Field

As it was already noted, standard definitions (see [40, 42, 54, 55]) of macroscopic field vectors **D** and **H** describing the medium response to the mean electromagnetic field (**E**, **B**)

$$\mathbf{D} \equiv \varepsilon_h \mathbf{E} + \mathbf{P}, \qquad \mathbf{H} \equiv (\mu_0)^{-1}\mathbf{B} - \mathbf{M} \qquad (3.1)$$

are adopted in phenomenological models. Here **P** and **M** are (as above) the electric and magnetic dipole moments per unit volume of the medium, respectively, and $\varepsilon_h = \varepsilon_0 \varepsilon_m$ is absolute permittivity of the host medium (for composites $\varepsilon_m \ne 1$).

In this formalism, which initially belongs to H. A. Lorentz [74], mean field vectors **E** and **B** are claimed to be main macroscopic field vectors because their physical meaning corresponds to the formula for the Lorentz force: $\mathbf{F} = q\mathbf{E} + q\mathbf{v} \times \mathbf{B}$. Here q is the probe charge, moving in the medium with the speed $\mathbf{v}$. The probe charge is by definition small enough so that to neglect the medium polarization by it. Respectively, the pair $(\mathbf{D}, \mathbf{H})$ is auxiliary and their physical meaning is exhausted by their definitions.

However, the claim that macroscopic values $\mathbf{E}$ and $\mathbf{B}$ must be obviously primary field vectors is disputable. In fact, a probe charge (such as a free electron in the medium) must be microscopically small. Therefore, it feels the microscopic (true) field and not the mean field. The argument by Lorentz that mean fields $\mathbf{E}$ and $\mathbf{B}$ are primary because they result from the averaging of truly primary fields $\mathbf{E}^{\text{true}}$ and $\mathbf{B}^{\text{true}}$ acting on a free electron is hardly convincing. In fact, the probe charge must move in between the medium particles, otherwise it will be adsorbed. However, the host medium for particles of the natural medium is free space where $\mathbf{B}^{\text{true}} = \mu_0 \mathbf{H}^{\text{true}}$. For a probe charge there is no difference between $\mathbf{B}^{\text{true}}$ and $\mathbf{H}^{\text{true}}$.

Therefore, the introduction of primary field vectors based on the arguing of the probe charge is disputable. An alternative phenomenology (see [41, 52, 53, 80])

$$\mathbf{D} \equiv \varepsilon_h \mathbf{E} + \mathbf{P}, \quad \mathbf{B} \equiv \mu_0 \mathbf{H} + \mathbf{M} \qquad (3.2)$$

, where the primary mean magnetic field is $\mathbf{H}$ and the magnetic medium response is attributed to $\mathbf{B}$: is not worse than the Lorentz phenomenology. Notice that in the second formalism, the definition of the magnetic moment differs by a factor μ_0. This is the only difference between these models.

The key point for us is not the choice of primary vectors but the fact that the polarization in any particle is induced by the local electric field varying in time and space. Action of the magnetic field vector $\mathbf{B}^{\text{loc}}$ in our case, when the magnetism is artificial and the bianisotropy is reciprocal, is not distinguishable from the action of the non-potential part of spatially non-uniform $\mathbf{E}^{\text{loc}}$.

In this Chapter, we have postulated that the macroscopic electric field $\mathbf{E}$ is uniquely related with the local field $\mathbf{E}^{\text{loc}}$ at centers of every particle. Therefore, the spatially non-uniform $\mathbf{E}^{\text{loc}}$ can be presented as the action of spatially non-uniform $\mathbf{E}$.

We may start from Maxwell's equations for time-harmonic mean fields:

$$\nabla\times\mathbf{E} = -j\omega\mathbf{B}, \quad \nabla\cdot\mathbf{B} = 0, \qquad (3.3)$$

and write for the spatial distribution of the macroscopic density of charges ρ

$$\nabla\cdot\mathbf{E} = \frac{\rho}{\varepsilon_h}. \qquad (3.4)$$

Finally, for the spatial distribution of the induced density of currents **J** we have

$$\mu_0\mathbf{J} = \nabla\times\mathbf{B} - j\omega\varepsilon_h\mu_0\mathbf{E}. \qquad (3.5)$$

Here macroscopic **J** and ρ are linked by the continuity equation:

$$\nabla\cdot\mathbf{J} = -j\omega\rho, \qquad (3.6)$$

Notice that Eq. (3.5) corresponds to the Lorentzian formalism. An alternative form of this Maxwell's equation is

$$\mathbf{J} = \nabla\times\mathbf{H} - j\omega\varepsilon_h\mathbf{E}. \qquad (3.7)$$

Equation (3.7) assumes that the primary mean magnetic field is **H**. This approach results in the same material equations as we will derive below. Since this Chapter follows the original papers and book [1], where the Lorentz formalism was chosen, below we use Maxwell's equation (3.5).

Since our homogenization model is not phenomenological, we have no right to simply define auxiliary mean field vectors **D** and **H** by Eqs. (3.1) or (3.2). We have to define them so that to get rid of the macroscopic polarization currents and charges in Maxwell's equations attributing the whole polarization/magnetization response to **D** and **H**. Therefore, we define these vectors rewriting Eqs. (3.4) and (3.5) in the form:

$$\nabla\times\mathbf{H} = j\omega\mathbf{D}, \qquad (3.8)$$

$$\nabla\cdot\mathbf{D} = 0, \qquad (3.9)$$

This means that auxiliary vectors **D** and **H** are defined via **E** and **B** and some vectors $\mathbf{T}_e$ and $\mathbf{T}_m$, describing the medium response:

$$\mathbf{D} = \varepsilon_h\mathbf{E} + \mathbf{T}_e, \qquad (3.10)$$

$$\mathbf{H} = \mu_0^{-1}\mathbf{B} + \mathbf{T}_m, \quad (3.11)$$

that we have to find.

Equating relations (3.8) and (3.9), where (3.10) and (3.11) are substituted, to initial relations (3.5) and (3.6) we obtain:

$$\nabla \times \mathbf{T}_m = j\omega \mathbf{T}_e - \mathbf{J}, \quad (3.12)$$

$$\nabla \cdot \mathbf{T}_e = -\rho. \quad (3.13)$$

Equations (3.12) and (3.13) offer a freedom to auxiliary electromagnetic vectors $\mathbf{D}$ and $\mathbf{H}$ granted by the vector identity $\nabla \cdot \nabla \times \mathbf{F}(\mathbf{r}) = 0$. This freedom will be used below twice. First, we will use it in order to share the magnetic medium response. Second, we will use it in order to derive the material equations so that they will be valid everywhere until the effective medium surface and even outside the medium. This covariance of MEs, as we will see, is the prerequisite of the compatibility of the model with Maxwell's boundary conditions.

At the present stage we simply choose $\mathbf{T}_e = \mathbf{J}/j\omega$ so that to nullify the right-hand side of (3.12). Then we have:

$$\mathbf{D} = \varepsilon_h \mathbf{E} + \frac{\mathbf{J}}{j\omega}, \quad (3.14)$$

$$\nabla \times \mathbf{T}_m = 0, \quad (3.15)$$

and our vector $\mathbf{T}_m$ turns out to be potential.

This means that defining $\mathbf{D}$ by Eq. (3.14), we introduce a potential medium response described by a corresponding vector $\mathbf{T}_m$ into $\mathbf{H}$ defined by Eq. (3.11). This part of $\mathbf{H}$ is not physically sound— magnetic phenomena are non-potential ones. Therefore, at this stage $\mathbf{T}_m = 0$, and vector $\mathbf{H}$ is related to $\mathbf{B}$ in the same way as in free space:

$$\mathbf{H} = \mu_0^{-1}\mathbf{B}. \quad (3.16)$$

Such a definition results in the trivial relative permeability $\mu = 1$, i.e., at this stage we have attributed all magnetic phenomena to the electric displacement vector $\mathbf{D}$. It is clear in advance that this $\mathbf{D}$ cannot be compatible with Maxwell's boundary conditions and the relations we will derive at this stage will be not MEs. Also, the definition of $\mathbf{H}$ expressed by Eq. (3.16) does not allow us to match the tangential component of **H** in the medium with its analogue in free space at the medium interface.

Fortunately, we may use the freedom in $\mathbf{T}_e$ and $\mathbf{T}_m$ and replace our initial vectors **D** and **H** defined by (3.14) and (3.16) to those compatible with Maxwell's boundary conditions. And we will do it in the next stage.

3.3 Multipole Expansion of the Macroscopic Polarization Current

3.3.1 Microscopic and Macroscopic Multipole Densities

In order to correctly replace the initial unsuitable definitions (3.14) and (3.16) by suitable definitions and make new medium-response vectors $\mathbf{T}_e$ and $\mathbf{T}_m$ be physically sound we have to analyze how macroscopic current **J** is related to the mean field **E**. This analysis will result in quasi-material equations of a medium with WSD and is based on the multipole expansion of the mean polarization current **J**.

Before we introduce this multipole expansion, let us make some preliminary remarks on how the mean polarization current **J** is obtained from the microscopic polarization response. To imagine this procedure let us mentally split the array of "molecules" (natural or artificial, it does not matter) onto unit cells $d \times d \times d$, so that each of them includes one molecule as it is shown in Fig. 3.1. We have to express **J** through **E** assuming that we know how the microscopic polarization currents $\mathbf{J}^{\text{micro}}$ inside the particles are related to the local electric fields $\mathbf{E}^{\text{loc}}$ somehow distributed over the particles.

It would be erroneous to introduce macroscopic current **J** by a simple averaging of $\mathbf{J}^{\text{micro}}$ similar to (2.12):

$$\mathbf{J}(\mathbf{r}) \neq \frac{1}{V} \int_V \mathbf{J}^{\text{micro}}(\mathbf{r} - \mathbf{r}') \, d^3\mathbf{r}'.$$

True electromagnetic field ($\mathbf{E}^{\text{true}}$, $\mathbf{H}^{\text{true}}$) originates from both potential and curl components of $\mathbf{J}^{\text{micro}}$. If we spread ($\mathbf{E}^{\text{true}}$, $\mathbf{H}^{\text{true}}$) by a similar bulk averaging we will not lose the contribution of the curl microscopic currents because d is small compared to the wavelength and the information about the eigenwave is not lost. However, the curls of $\mathbf{J}^{\text{micro}}$ occur within a particle and the integration over the unit cell nullifies them and the curl component in **J** will be absent. Such a model will be not self-consistent. Averaging of the microscopic polarization current is not straightforward, and the only known practical way to this averaging is multipole decomposition of both microscopic and macroscopic currents.

Not all researchers are happy with the multipole description of the medium response. Some researchers believe that the description of a particle by a set of multipoles referred to a certain point of the unit cell is physically not adequate. They think that the replacement of particles but p-dipoles or even by sets of multipoles is the same as negligence of their physical sizes and implies the non-physical infinite values for the microscopic fields calculated at the centers of the particles.

Of course, the replacement of particles forming a medium with WSD by point scatterers is not physically adequate. In the model by Madelung, Ewald and Oseen this replacement resulted in the singular microscopic fields at the dipole centers. This approach is not self-consistent and resulted in the correct model because Ewald and Oseen averaged the true fields in terms of *principal values*. Only implying V.P. in the integral (2.12) the model of point dipoles allows us to avoid the singularities at the dipole centers. On the contrary, our homogenization model does not replace the particle by a point scatterer. When we speak on the centers of constitutive particles, we only mean that the multipole moments are referred to these points. This is necessary to specify this point **R** for every particle because multipole moments of the same microscopic polarization current distribution (all of them but the p-dipole) depend on the coordinates of the points to which they refer. However, we do not replace the finite density of these multipole moments by delta-functions at **R**. Microscopic multipole densities are thought to be uniformly distributed over the physical volume of the particle V_p. And the bulk averaging of these microscopic multipole densities over the unit cell volume V gives the same result as the simple division of the particle multipole moments by V.

As to the electromagnetic interaction is described by the CMLL formula (2.4) which will be specially revised in the next chapter for finite-size particles in the dynamic case. Thus, our description of the

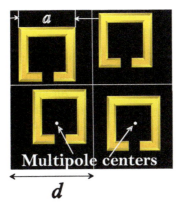

Figure 3.1 Illustration to the concept of the medium unit cell. Multipoles of the constitutive particles are referred to the centers of corresponding unit cells. In the irregular case, the multipole moments at the cell centers will be different even for the same mean field in different cells. However, this difference does not cause a trouble if our MEs are covariant.

particle response in terms of microscopic multipoles is fully self-consistent with the homogenization model.

In the array illustrated by Fig. 3.1 the orientations of all particles is identical. To conclude this subsection, let us discuss the impact of deviation in the orientations of the constitutive particles. The averaging procedure for the arrays with partially or fully random orientations of particles is very simple. If particles are oriented fully randomly, the microscopic multipole polarizabilities of a single particle are averaged so that the tensors degenerate into scalar values. These scalars are mean values of the diagonal components of the corresponding microscopic tensors. Off-diagonal components of these tensors nullify. Thus, the array of randomly oriented particles is replaced by an array of isotropic multipolar scatterers. Their bulk averaging will result in the set of scalar EMPs (such as permittivity, permeability, and MEC parameter). If the deviation in the particle orientations is not fully random, one averages the microscopic multipole polarizabilities over the corresponding sheer of particles orientations. Then the averaged effective particle will be still anisotropic but its anisotropy will be lower than that of an original constitutive particle.

3.3.2 Attempts to Avoid Multipoles in the Model of Weak Spatial Dispersion

Some researchers claim that multipoles are cumbersome and difficult. This may be really so if we have no sufficient experience with tensors. However, what to do? All known attempts to avoid the multipole decomposition for media with WSD resulted in some strange homogenization models which are either incomplete or even more difficult than the multipole one.

In [149] it was suggested (and in [106] it was done) to average the polarization current $\mathbf{J}^{\text{micro}}$ by integration with two weight functions f_e^{av} and f_m^{av}. Averaging of the same microscopic distribution of the polarization current with one weight function results in the electric polarization and that with another weight function results in the magnetization. However, there is no theory on how to properly choose these weight functions. No receipt for composite media are known that would make this approach practically working.

Although in work [106] an attempt to suggest $f_{e,m}^{av}$ for some molecular media with WSD was done, it did not result in any practical homogenization model because the macroscopic fields corresponding to these exotic polarization and magnetization are not mean fields and do not obey to usual Maxwell equations for mean fields. These exotic macroscopic fields were called in [106] *finite fields*. How to apply finite fields for solving the boundary problems is unclear.

The second attempt to avoid multipoles in media with WSD was done in work [160]. Its authors introduce original definitions of mean fields and currents for lattices with artificial magnetism and bianisotropy. The definition of the mean field via the true field is absent as such. As to averaging the microscopic polarization currents, the authors replace a constitutive particle by a set of discrete point dipoles. A unit cell now contains a set of point dipoles. The next step is a smooth extrapolation of this discrete polarization to the unit cell volume. This spread dipole polarization $\mathbf{P}$ is maximal inside the particle and outside it $\mathbf{P}$ decays but keeps nonzero. Next, the spread polarization current $\mathbf{J} = j\omega\mathbf{P}$ is introduced. Since this current is continuous, the effective medium is continuous automatically. Such a homogenization model does not distinguish the microscopic and macroscopic parameters. Particles are considered as sources of the electromagnetic field in the structure and after their meshing are replaced by the spread sources.

In this model, the mean electric and magnetic field of the eigenmode are found as those produces by these spread impressed sources **P** and **J**:

$$\nabla\times\nabla\times\mathbf{E} - k^2\mathbf{E} = \mu_0\mathbf{P}, \quad \nabla\times\nabla\times\mathbf{H} - k^2\mathbf{H} = \nabla\times\mathbf{J},$$

where k is the host matrix wave number. The solutions of these two equations are obtained in [160] using the volume integral equation derived for a unit cell from the Bloch periodical conditions on the opposite faces of the cell. As it is known from the theory of crystals Bloch periodical conditions imply the knowledge of the effective wave vector **q**. This solution is then interpreted as the solution of the cell problem in terms of the non-local effective permittivity. The last one for the given **q** and ω is found by a fitting procedure. A special fitting allows one to extract the permeability (second-order term with respect to **q**) and MEC parameter (first-order term) from this non-local permittivity. In this way one may transit from the non-local permittivity to the local one, and also to the local permeability and MEC parameter. Below we will see how to do it explicitly analyzing the SD of the second order.

Thus, work [160] suggests a numerical homogenization model— an algorithm for calculating EMPs based on the assumption that the array can be described as a BA medium with artificial magnetism. The refusal of the multipole resulted in the absence of an analytical homogenization model and even in postulated material equations instead of derived ones. It only aims the eigenmode and it remains unclear can these local EMPs be used for solving boundary problems or not.

An attempt to reduce the amount of involved multipoles was done in the already mentioned book [63]. Here, the idea of [160] was developed and extended to irregular densely packed composites of complex-shape metal particles. The homogenization model of [63] consists of two stages. First, one shares out a rather large cubic domain of the array (of the order of the effective wavelength) called supercell. Since the supercell comprises enough particles, its response comprises all features of the effective medium. Therefore, all properties of the original random array are represented by one internally random supercell.

A supercell is meshed so that the mesh unit cell is much smaller than the particle size and each unit cell is modeled as a dipole. One numerically solves the boundary problem for the supercell impinged by a plane wave and calculates the microscopic polarization in the mesh nodes. This calculation results in a discrete set of p-dipoles (or values of the microscopic polarization current) replacing the original structure in the supercell. Up to this point the method is similar to that of [160].

However, the second stage of the model is different. Instead of spreading these polarizations, one introduces a larger mesh for the supercell. A new unit cell contains a cubic group of several (e.g., eight) p-dipoles that is sufficiently large to be described by three multipoles—a total p-dipole of the new unit cell, its m-dipole, and its electric quadrupole. After calculating these three multipoles, one relates them with the electric and magnetic fields averaged over new unit cells. This way, one finds three local susceptibilities that still can be different for different unit cells. If they are different, this means that the medium model still comprises the subwavelength fluctuations. Then one increase the unit cell and repeats the calculation of the p-dipole, m-dipole and electric quadrupole susceptibilities until they stop to vary from cell to cell. The final step is attributing these three susceptibilities to the supercell. In this way, the supercell is modeled by a continuous medium with artificial magnetism and bianisotropy, and its three EMPs—permittivity, permeability, and the MEC parameter—are found through the numerical simulation of a supercell.

We see that this approach is another algorithm of numerical homogenization. It is applicable not only to periodic structures, however, its weak point is its application to boundary problems, where the medium sample and the wave incidence may be different from those adopted in the initial simulations. This is so because this numerical model is still a heuristic description and may be not compatible with Maxwell's boundary conditions (it was honestly pointed out in [63]).

3.3.3 On the Advantages of Multipoles

All known attempts to avoid multipoles in a complete homogenization model of media with WSD resulted in something more difficult and less practical. Meanwhile, the multipole homogenization has some evident advantages. This homogenization model is the same for regular and slightly irregular arrangement of particles like that shown in Fig. 3.1. The equivalence of the mean electromagnetic response of a slightly irregular optically dense array to a regular lattice will be seen from the covariant form of our MEs. Namely, we will show that our EMPs do not depend on the exact location of the multipole center of each particle. An only condition of the same homogenization model for both regular and irregular cases is one particle per one unit cell.

Notice that an alternative explanation of the equivalence of a slightly random array to a regular one is given in work [162]. The logics of [162] is as follows. Consider a regular lattice of p-dipole particles with period d, calculate the local field acting on any particle in it (via the dynamic dipole sums) and the induced dipole moment. Then let us introduce the slight irregularity shifting the reference particle by a small vector $\mathbf{t}$ ($t < d/2$) from the lattice node. Since this displacement is smaller than $d/2$ the particle remains in the same unit cell. Then the local field $\mathbf{E}^{loc}$ acting on the reference particle and, consequently, its local polarization $\mathbf{P}$ will change so that $\mathbf{P}/3\varepsilon_0$ in the CMLL formulas. The difference $\mathbf{E}^{loc} - \mathbf{P}/3\varepsilon_0$ that is equal to the mean field keeps the same.[b] So, a small displacement of the reference particle does not change the mean field in the reference unit cell. Therefore, the mean field of the whole array is not sensitive to the small derivations in its particle positions.

An advantage of the multipole theory is that the equivalence of the regular and irregular structures is obtained automatically. However, the mean advantage is the possibility to build a complete homogenization model. For the first time, this was noticed in works [38, 85, 87, 161] published prior to [1]. However, in these works this homogenization model was not built. In order to stress the scientific

---

[b]In [162] it was assumed that the mean field E may change due to the particle shift but returns to its regular value after a statistic averaging over all possible vectors t. However, it can be shown that E is invariant to a so small shift that $t < d/2$.

novelty of the theory developed in [1] let us specify the contribution of [38, 85, 87, 161] into the theory of WSD. It was as follows:

- Expressions relating **J** and **E** were derived and the hierarchy of multipole moments was explained on condition $(d/\lambda_{\text{eff}}) \ll 1$. This hierarchy means that multipoles of the n-th order are smaller than the multipoles of the $n-1$-th order. The hierarchy starts from the p-dipole moment having the zeroth order of smallness. The first-order multipoles are the m-dipole moment and the electric quadrupole moment. The second-order multipoles are magnetic quadrupole and electric octupole, etc. (in each order of smallness starting from the first one there is one magnetic and one electric multipole).
- It was shown: if this hierarchy is respected, the series of multipoles converges and in practice can be truncated.
- Analyzing this series the terms responsible for artificial magnetism (second order) and bianisotropy (first order) were shared out.
- QMEs were derived and their non-covariance was noticed.
- Covariant MEs were derived taking into account the terms of the first order and the necessity to do it in the second order was emphasized.

In the present book these earlier results are merged with the original ones so that to present the complete quasi-static homogenization model of media with WSD in a logical way suitable for educational purposes.

First, we derive the MEs and determine the EMPs for any optically dense array of particles and after that we build the homogenization model relating these EMPs with the microscopic response of an individual particle. In this section we analyze the properties of the macroscopic multipole polarization assuming that the microscopic multipole densities are already spread in the array volume.

3.3.4 Multipole Moments

As it was already mentioned, Cartesian multipoles split onto two groups—electric and magnetic multipoles. Both these types exist in each order of smallness except the zero one. Magnetic multipoles reflect the curvature of the polarization current lines and the out-of-phase distribution of the polarization current oscillations. Only the magnetic dipole can survive in absence of oscillations. Artificial magnetism means that even the magnetic dipole vanishes in the static limit. Electric multipoles reflect the density of the polarization current lines and the non-uniformity of this density. No one of electric multipoles vanishes in the static limit.

Multipole moments of a particle with volume V_p are equally defined through microscopic charge density ρ^{micro} or density of microscopic polarization current $\mathbf{J}^{micro}$ [87]. The equivalence is ensured by the continuity equation. In order to write these definitions for the time-harmonic case the most suitable way is to use the index form of tensors:

- Electric dipole moment is a vector with components:

$$p_\alpha = \int_{V_p} \rho^{micro} r_\alpha \, d^3\mathbf{r} = \frac{1}{j\omega} \int_{V_p} J_\alpha^{micro} \, d^3\mathbf{r}, \quad (3.17)$$

- Electric quadrupole moment is a dyad with components:

$$q_{\alpha\beta} = \int_{V_p} \rho^{micro} r_\alpha r_\beta \, d^3\mathbf{r} = \frac{1}{j\omega} \int_{V_p} (J_\alpha^{micro} r_\beta + J_\beta^{micro} r_\alpha) \, d^3\mathbf{r}, \quad (3.18)$$

- Electric octupole moment is a triad with components:

$$o_{\alpha\beta\gamma} = \int_{V_p} \rho^{micro} r_\alpha r_\beta r_\gamma \, d^3\mathbf{r}$$

$$= \frac{1}{j\omega} \int_V (J_\alpha^{micro} r_\beta r_\gamma + J_\beta^{micro} r_\alpha r_\gamma + J_\gamma^{micro} r_\beta r_\alpha) \, d^3\mathbf{r}, \quad (3.19)$$

- Magnetic dipole moment is a vector with components:

$$m_\alpha = \frac{1}{2} \int_{V_p} (r_\beta J_\gamma^{micro} - r_\gamma J_\beta^{micro}) \, d^3\mathbf{r} = \frac{1}{2} \int_{V_{micro}} [\mathbf{r} \times \mathbf{J}^{micro}]_\alpha \, d^3\mathbf{r}, \quad (3.20)$$

- Magnetic quadrupole moment is a dyad with components:

$$S_{\alpha\beta} = \frac{1}{3} \int_{V_{micro}} ([\mathbf{r} \times \mathbf{J}^{micro}]_\beta r_\alpha \, d^3\mathbf{r}, \qquad (3.21)$$

The definitions of more multipole moments can be found in [87]. Radius-vector $\mathbf{r}$ in these formulas originates from the selected center of the particle $\mathbf{r} = 0$. It is an arbitrary point that may be slightly shifted even outside the particle so that its distance to any points of the particle is optically small. Calculating the particle multipoles we assume them to be referred to the point $\mathbf{r} = 0$. However, the particle is thought as a scatterer with a microscopic density of multipole moments uniformly distributed in its volume V_p. No one of these multipoles creates the infinite field at $\mathbf{r} = 0$. However, it is worthy to note that after the calculation of multipoles we are not interested in finding the correct microscopic fields. All we need is the mean field and the densities of the mean multipole moments. These densities inside the given unit cell are equal to the microscopic multipole moments of the given particle divided by the unit cell volume.

Now, let us discuss again the choice of the multipole center. For a spherical particle this choice is evident. However, if the spherical particles form a slightly irregular array, nothing forbids us to refer the multipole centers of different spheres to the points shifted from their centers so that all multipole centers were centers of the unit cells as shown in Fig. 3.1. This will not change the homogenization model if the EMPs are covariant—do not depend on the choice of $\mathbf{r} = 0$ within the unit cell. This point of our theory is especially important for scatterers whose geometry does not offer an evident choice of the multipole center. In fact, an effective particle can a non-symmetric dimer like that shown in Fig. 2.3e or even an oligomer. For such particles the model must be evidently non-sensitive to the choice of the point $\mathbf{r} = 0$.

In the next subsection we have in mind the following smooth functions of coordinates: $\mathbf{P}$—density of the medium p-dipole moments, $\bar{\bar{Q}}$—density of the medium electric quadrupole moments, $\mathbf{M}$—density of the medium m-dipole moments, etc. These multipole densities vary from one unit cell to another unit cell as it is imposed by the wave propagating in the original array. If the array is a finite-thickness layer, there are also standing waves formed by internal reflections at the interfaces. Then the spatial variations of the multipole densities will correspond to these standing waves. For the multipole susceptibilities of our effective medium (we aim to analyze) it does not matter because we derive them so that they do not depend on the wave vector.

3.3.5 Polarization Current in Media with Weak Spatial Dispersion

Both mean field **E** and mean polarization current **J** vary from cell to cell as spatially smooth functions. Their variations in the spatial scale of a unit cell d and in that of a particle a are not negligible. Microscopic multipoles within the particle are determined by the local electric field distributed somehow over the particle. Macroscopic multipole densities are spread microscopic multipoles and since the local and mean field are uniquely related, they can be related to the non-uniform mean electric field. Macroscopic polarization current **J(R)** that we will present here via macroscopic multipole densities is linked with the mean electric field **E** distributed in a certain spatial domain Ω around **R**. We have already discussed this effective volume introducing the general relation (1.1) for **J** in the introductory chapter.

Now, let us write this relation in a more strict form—using the tensor notations:

$$J_i(\mathbf{R}) = \int_\Omega K_{ij}(\mathbf{R} - \mathbf{R}') E_j(\mathbf{R}') \, dV'. \qquad (3.22)$$

Here K_{ij} are components of the polarization current response dyad (indices i and j denote here the Cartesian components). Below, inspecting the applicability of the CMLL formulas to media with WSD we will estimate the domain Ω and see that it is simply equal to the volume $V = d^3$ of the unit cell of the original array. Therefore, in (3.22) we may have in mind that $|\mathbf{R} - \mathbf{R}'| \leq d$.

The electric field at any point inside Ω can be expanded into the Taylor series around the center:

$$\mathbf{E}(\mathbf{R}') = \mathbf{E}(\mathbf{R}) + (\nabla_\alpha \mathbf{E})\Big|_\mathbf{R} (R'_\alpha - R_\alpha) + \frac{1}{2}(\nabla_\beta \nabla_\alpha \mathbf{E})\Big|_\mathbf{R}$$
$$\times (R'_\alpha - R_\alpha)(R'_\beta - R_\beta) + \cdots \qquad (3.23)$$

The substitution of (3.23) into (3.22) leads to the Taylor expansion of the polarization current (1.3) that can be written in the form:

$$J_i(\mathbf{R}) = b_{ij} E_j(\mathbf{R}) + b'_{ijk} \nabla_k E_j(\mathbf{R}) + b''_{ijkl} \nabla_l \nabla_k E_j(\mathbf{R}) + \cdots \quad (3.24)$$

Here b_{ij}, b'_{ijk}, etc., denote different tensors resulting from the integration of functions $K_{ij}(\mathbf{r})$, $K_{ij}(\mathbf{r})r_k$, etc., around the observation point $\mathbf{R}$ ($\mathbf{r} \equiv \mathbf{R} - \mathbf{R}'$). In the notations of (1.3) $b_{ij} = j\omega \kappa_{ij}$, $b_{ij} = j\omega \kappa'_{ijk}$, etc.

The expression (3.24) describes the phenomenon of WSD in terms of the averaged polarization current $\mathbf{J}(\mathbf{R})$ and is covariant—does not depend on the choice of the multipole center in the unit cell Ω. We have already obtained (3.24) in the introductory chapter, where it was written in the simplified form. Now we write it with the tensor notations for the further use in our derivations.

Three points following from the expansions (3.22)–(3.24) deserve to be stressed:

- In (3.22) it is assumed that natural magnetic inclusions are absent.
- The action of the time-dependent magnetic field to nonmagnetic inclusions is taken into account by formula (3.22)—it is the response of the medium unit cell to the solenoidal part of the spatially varying local electric field.
- In Eq. (3.24) we use the mean field $\mathbf{E}$, though, in fact, the polarization current is induced by the local field. However, $\mathbf{E}_{\text{loc}}$ is assumed to be uniquely determined by two zero-order values—mean field $\mathbf{E}$ and dipole polarization $\mathbf{P}$. Then the the replacement of the local field by the mean field keeps same orders of smallness to all terms of the expansion (3.24).

3.3.6 Electric and Magnetic Polarization Currents

In media with multipole polarization, the polarization current can be presented through the spatial derivatives of the macroscopic multipole moment densities. For microscopic polarization current and microscopic multipole densities (an individual scatterer) this expansion was derived in works [38, 87]. In [42, 56, 85] it was expanded to the media—the proof is the same, the difference is only the spatial scale. We will not spend time for reproduction of this expansion, it is enough to notice that it follows from (3.22) and tightly related with (3.24). In the index form this expansion is as follows:

$$J_\alpha = j\omega P_\alpha - \frac{j\omega}{2}\nabla_\beta Q_{\alpha\beta} + e_{\alpha\beta\gamma}\nabla_\beta M_\gamma + \frac{j\omega}{6}\nabla_\gamma \nabla_\beta O_{\alpha\beta\gamma}$$
$$- \frac{1}{2}e_{\alpha\beta\delta}\nabla_\gamma \nabla_\beta S_{\delta\gamma} + \cdots \quad (3.25)$$

Here P_α are Cartesian components of the electric dipole polarization vector (Greek or Latin indices below denote the coordinate axes x, y, z), $Q_{\alpha\beta}$ are the components of the electric quadrupole polarization tensor (dyadic), M_γ are the components of the magnetic dipole polarization vector, $S_{\delta\gamma}$ are the components of the magnetic quadrupole polarization tensor (dyadic), and $O_{\alpha\beta\gamma}$ are the components of the electric octopole polarization tensor (triadic). The Levi–Civita tensor $\overline{\overline{\overline{e}}}$ is an antisymmetric unit triad with nonzero off-diagonal component components $e_{xyz,zxy,yzx} = 1$ and $e_{xzy,zyx,yxz} = -1$ (other components are zeros). This anti-symmetric triad, by the way, allows one to present the curl operator acting on an arbitrary vector field **N** as follows:

$$(\nabla \times \mathbf{N})_\alpha = e_{\alpha\beta\gamma}\nabla_\alpha N_\gamma. \quad (3.26)$$

Therefore, expansion (3.25) can be rewritten in a more compact tensor form:

$$\mathbf{J} = j\omega\mathbf{P} - \frac{j\omega}{2}\nabla\overline{\overline{Q}} + \nabla\times\mathbf{M} + \frac{j\omega}{6}\nabla\nabla\overline{\overline{Q}} - \frac{1}{2}\nabla\times\nabla\overline{\overline{S}}. \quad (3.27)$$

Expansion (3.27) can be found also in popular books [42, 53]. However, the notations used for multipole moments in these books are more complicated, and we will use simple and transparent notations of works [38, 56, 85, 87].

Taking into account the multipoles explicitly written on the right-hand side of (3.27), we take into account all effects of both first-order spatial dispersion (bianisotropy) and second-order spatial dispersion (such artificial magnetism). Higher-order multipoles neglected in (3.25) do not correspond to any important physical effects, as we have already noticed.

We can see that the polarization current is the sum of two components: One of them is the vortex-free (potential) polarization current that is often called the electrical one, and the other one is the vortex-type (non-potential) polarization current, that is often called the magnetic one:

$$\mathbf{J} = \mathbf{J}^{el} + \nabla \times \mathbf{J}^{mag}, \tag{3.28}$$

$$\mathbf{J}^{el} = j\omega \mathbf{P} - \frac{j\omega}{2} \nabla \bar{\bar{Q}} + \cdots, \tag{3.29}$$

$$\mathbf{J}^{mag} = \mathbf{M} - \frac{1}{2} \nabla \bar{\bar{S}} + \cdots \tag{3.30}$$

Recall that we do not consider the *natural* magnetism and this magnetic current originates from the curl part of the microscopic electric polarization currents. Equations (3.28)–(3.30) together with the expansion (3.24) will be used below for the introduction of EMPs.

3.4 Material Equations for Media with Weak Spatial Dispersion

3.4.1 Non-Covariant Constitutive Equations

At this stage we do not distinguish ME and QME, yet. Such equations are usually introduced for time-harmonic fields in order to get rid of averaged polarization current and polarization charges (linked with one another by the continuity equation) in corresponding Maxwell's equations (3.3), (3.8), and (3.9) for macroscopic fields. Here we repeat these equations for the convenience of the reader:

$$\nabla \times \mathbf{E} = -j\omega \mathbf{B}, \tag{3.31}$$

$$\nabla \cdot \mathbf{B} = 0, \tag{3.32}$$

$$\nabla \times \mathbf{B} = j\omega \varepsilon_h \mu_0 \mathbf{E} + \mu_0 \mathbf{J}, \tag{3.33}$$

$$\nabla \cdot \mathbf{E} = \frac{\rho}{\varepsilon_h} = -\frac{\nabla \cdot \mathbf{J}}{j\omega \varepsilon_h}. \tag{3.34}$$

We have already explained that the vectors **D** and **H** are introduced so that to replace Maxwell's equations (3.33) and (3.34) by equations that do not contain the polarization currents and charges.

Substituting (3.28) into (3.34) we can see that in fact

$$\nabla \cdot \mathbf{E} = -\nabla \cdot \left(\frac{1}{j\omega\varepsilon_h} \mathbf{J}^{\text{el}} \right). \qquad (3.35)$$

Substituting (3.28) into (3.33) we obtain:

$$\nabla \times (\mathbf{B} - \mathbf{J}^{\text{mag}}) = j\omega\mu_0(\varepsilon_h \mathbf{E} + \mathbf{J}^{\text{el}}). \qquad (3.36)$$

Therefore, instead of defining **D** by Eq. (3.14) and **H** by Eq. (3.16) it is reasonable to define these auxiliary vectors as follows:

$$\mathbf{D} = \varepsilon_h \mathbf{E} + \mathbf{J}^{\text{el}}, \quad \mathbf{H} = \mu_0^{-1} \mathbf{B} - \mathbf{J}^{\text{mag}}. \qquad (3.37)$$

With these definitions we also obtain correct macroscopic Maxwell's equations

$$\nabla \times \mathbf{H} = j\omega \mathbf{D}, \qquad (3.38)$$

$$\nabla \cdot \mathbf{D} = 0, \qquad (3.39)$$

complementing initial macroscopic Maxwell's equations (3.31) and (3.32). Taking into account (3.29) and (3.30), one can see that vectors **D** and **H** are now defined through the multipole densities:

$$\mathbf{D} = \varepsilon_h \mathbf{E} + \mathbf{P} - \frac{1}{2}\nabla \cdot \overline{\overline{Q}} + \frac{1}{6}\nabla \cdot (\nabla \cdot \overline{\overline{\overline{O}}}) + \cdots, \qquad (3.40)$$

$$\mathbf{H} = (\mu_0)^{-1}\mathbf{B} + \mathbf{M} - \frac{1}{2}\nabla \cdot \overline{\overline{S}} + \cdots . \qquad (3.41)$$

Only if no quadrupoles and high-order multipoles are induced in the medium particles, these equations are reduced to the usual formalism

$$\mathbf{D} = \varepsilon_h \mathbf{E} + \mathbf{P}, \qquad \mathbf{H} = (\mu_0)^{-1}\mathbf{B} - \mathbf{M}, \qquad (3.42)$$

resulting in the standard definitions of the effective permittivity and permeability and often treated as general definitions of **D** and **B**.

However, the media from particles containing only p-dipole and m-dipole moments represent only a special case of media with WSD. Such composites can be arrays of high-permittivity spheres, arrays of canonical CPs, or OPs at sufficiently low frequencies. For complex-shape particles (complex-shape molecules, metal helices and hooks, S-shaped particles, etc.) quadrupoles and, perhaps, even high-order multipoles must be taken into account.

The next step after (1.4) and (3.41) is to express the multipole densities through the averaged electric field **E** and its spatial derivatives comparing (3.25) with (3.24). This way we will relate vectors **D** and **B** with **E** and its derivatives. Using the same logics as we used deriving Eq. (3.24) we can write the similar expansions for each of multipole densities:

$$P_\alpha(\mathbf{R}) = a_{\alpha\beta} E_\beta(\mathbf{R}) + \frac{1}{2} a'_{\alpha\beta\gamma} \nabla_\gamma E_\beta(\mathbf{R}) + \frac{1}{6} a''_{\alpha\beta\gamma\delta} \nabla_\gamma \nabla_\delta E_\beta(\mathbf{R}) + \cdots \quad (3.43)$$

for the p-dipole polarization,

$$Q_{\alpha\beta} = Q'_{\alpha\beta\gamma} E_\gamma + \frac{1}{2} Q''_{\alpha\beta\gamma\delta} \nabla_\delta E_\gamma + \cdots \quad (3.44)$$

for the electric quadrupole one,

$$M_\alpha = M'_{\alpha\beta} E_\beta + \frac{1}{2} M''_{\alpha\beta\gamma} \nabla_\gamma E_\beta + \cdots \quad (3.45)$$

for the m-dipole one,

$$S_{\alpha\beta} = S'_{\alpha\beta} E_\beta + \cdots \quad (3.46)$$

for the magnetic quadrupole one, and finally:

$$O_{\alpha\beta\gamma} = O'_{\alpha\beta\gamma\delta} E_\delta + \cdots \quad (3.47)$$

for the electric octopole polarization $\overline{\overline{\overline{O}}}$. The second-order theory of SD implies the terms we have kept in these relations should be taken into account and the omitted terms are negligibly small.

Before substituting relations (3.43)–(3.47) into (1.4) and (3.41) we have to share out in the triad $\overline{\overline{\overline{a}}}'$ (of the p-dipole susceptibility to the first-order derivatives of **E**) two following parts. First part is symmetric with respect to the two last indexes. Second part is anti-symmetric. It easy to check (see also [163]) that any triad anti-symmetric for two last indexes can be presented as a scalar product of the Levi–Civita fully antisymmetric triad by a certain dyad (denoted below as $2\overline{\overline{g}}/j\omega$). Therefore, we can write:

$$a'_{\alpha\beta\gamma} = (a'_{\alpha\beta\gamma})^{\text{symm.}} + (a'_{\alpha\beta\gamma})^{\text{nonsym.}} \equiv d_{\alpha\beta\gamma} + e_{\delta\beta\gamma} \frac{2g_{\alpha\delta}}{j\omega}. \quad (3.48)$$

Let us keep in mind that the new triad $\overline{\overline{\overline{d}}} \equiv (\overline{\overline{\overline{a}}}')^{\text{symm.}}$ is the symmetric part of the triad $\overline{\overline{\overline{a}}}'$ and the new dyad $\overline{\overline{g}}$ determines its anti-symmetric remainder. An experienced reader looking at (3.43) and (3.48) has already guessed that the dyad $\bar{\bar{g}}$ describes the bianisotropy.

Another important formula (see [163]) relates the quadrupole susceptibility to the mean electric field with the symmetric triad $\bar{\bar{\bar{d}}}$:

$$Q'_{\alpha\beta\gamma} = Q'_{\beta\alpha\gamma} = d_{\gamma\alpha\beta} = d_{\gamma\beta\alpha}. \tag{3.49}$$

For the moment we simply postulate this formula. We will prove it below.

Substituting relations (3.43)–(3.47) into (1.4) and (3.41), we can also use Maxwell's equation (3.31) rewritten in the index form:

$$B_\beta = \frac{1}{j\omega} e_{\beta\alpha\gamma} \nabla_\alpha E_\gamma. \tag{3.50}$$

After substitutions of new relations (3.48) and (3.49) into (1.4) and (3.41) we easily obtain the following relations:

$$D_\alpha = \varepsilon_{\alpha\beta} E_\beta - g_{\alpha\beta} B_\beta + \frac{1}{2}(Q'_{\beta\gamma\alpha} - Q'_{\alpha\gamma\beta})\nabla_\beta E_\gamma + \Theta_{\alpha\beta\gamma\delta}\nabla_\beta\nabla_\delta E_\delta \tag{3.51}$$

and

$$H_\alpha = \mu_0^{-1} B_\alpha - M'_{\alpha\beta} E_\beta + \Gamma_{\alpha\beta\gamma}\nabla_\beta E_\gamma. \tag{3.52}$$

In these equations we used the following notations:

$$\varepsilon_{\alpha\beta} = \varepsilon_h I_{\alpha\beta} + a_{\alpha\beta}, \quad \Theta_{\alpha\beta\gamma\delta} = \frac{1}{6}(a''_{\alpha\delta\beta\gamma} - Q''_{\alpha\beta\gamma\delta} - O'_{\alpha\gamma\delta\beta}), \tag{3.53}$$

and

$$\Gamma_{\alpha\beta\gamma} = \frac{1}{2}(M''_{\alpha\beta\gamma} - S'_{\alpha\beta\gamma}). \tag{3.54}$$

Here $I_{\alpha\beta}$ is unit planar dyad: $I_{\alpha\beta} = 1$ if $\alpha = \beta$ and $I_{\alpha\beta} = 0$ if $\alpha \neq \beta$. The first formula in (3.53) coincides with the usual definition of the absolute permittivity comprising the factor ε_0. So, in media with WSD tensor $\bar{\bar{\varepsilon}}$ is still defined in the standard way—through the p-dipole susceptibility to the mean field. It does not include high-order multipole polarizabilites. The p-dipole susceptibility is the zero-order value in the initial expansion (3.24), and our effective permittivity purely corresponds to the zero-order response of the medium. Of course, this response can be dispersive and resonant. In the wave field $a_{\alpha\beta}$ can depend on the frequency as well as the dipole polarizability of a resonant particle. Words "zero order" refer only to SD, and not to frequency dispersion.

Equations (3.51) and (3.52) were obtained by R. Raab with colleagues in work [163] and in that work they were claimed to be general MEs of media with WSD (this wrong claim was corrected in [164] and in [85] their analogues covariant in the first order of smallness were obtained). As we have already discussed, these equations cannot be called MEs, because the dyad M'_{ij} and the triad $(Q'_{jki} - Q'_{ikj})$ are not covariant.

In [164] the dependence of these tensors on the location of the multipole center was analyzed for an individual particle. This origin dependence means that the magnetic dipole, and electric quadrupole (in fact, also magnetic quadrupole, electric octopole, and all higher multipoles) if taken separately from one another, are not measurable physical values.

Let us discuss the most important of these multipoles—the m-dipole. Only in two special cases the magnetic dipole susceptibility to the electric field M'_{ij} is covariant [56]. The first case corresponds to the frequencies at which the polarization current induced in the particle flows along a closed path (loop) and its density is uniform along this effective loop. The second case corresponds to the frequencies at which the p-dipole of the particle vanishes. The medium corresponding to the first case can be fabricated from particles performed as loops, e.g., from SRRs with very small splits. For such media the m-dipoles are covariant in a broad range of sufficiently low frequencies. The media in which the second regime is possible (e.g., an array of plasmonic dimers at a specific frequency) will be discussed below.

If the open part of a split ring in a SRR, improperly designed canonical CP or OP is comparable with the circumference of the wire loop, the magnetic dipole of the particle is not covariant. It is also not covariant for helices. Therefore, the magnetic dipole response of such particles cannot be exactly measured.

Which response is identified with the m-dipole in numerous works dedicated to such scatterers, e.g., in papers [166–168]? In these papers it is a combination of the magnetic dipole with the some higher multipoles, at least, with the electric quadrupole. Such a combination is covariant [164] and can be measured.

The dependence of the microscopic multipole moments of a molecule on the coordinates of the arbitrary chosen molecular center results in the dependence of averaged multipole moments (of the effective medium) on the Cartesian coordinate origin. If we move our coordinate frame, each multipole susceptibility changes. In fact, the mean field and its spatial derivatives do not depend on the position of the coordinate origin, but the multipole moments do. Therefore, our effective-medium susceptibilities and their combinations entering Eqs. (3.51) and (3.52), namely, M'_{ij} and $(Q'_{jki} - Q'_{ikj})$ are, in the general case, not measurable values. This fact obviously means that the higher multipoles cannot be neglected for composites of SRRs with large open parts considered in [166–168]. In these works the negligence of this fact resulted in a serious misinterpretation of the experimental and numerical data. Also, the higher multipoles are key importance for media of complex molecules such as organic dyes (see, e.g., [38, 85, 86]).

Physically sound EMPs of a medium cannot depend on the choice of the coordinate origin. This means that tensors M'_{ij} and $(Q'_{kij} - Q'_{kji})$ are not EMPs and can be called *non-covariant constitutive parameters*. Respectively, in presence of higher multipoles vectors **D** and **H** defined by two equations (3.37) are not physically sound electric displacement and magnetic tension vectors. In the next subsection, we will show in detail that such **D** and **B** do not allow us to solve boundary problems for an effective medium.

3.4.2 Why Non-Covariant Equations Cannot Be Used in Boundary Problems

Equations (3.51) and (3.52) were derived from the definitions of the auxiliary vectors of the mean field **D** and **H** which assume that the polarization of the medium is performed by the mean field **E**. In fact, it is performed by the local field. As we have discussed above, postulating the CMLL formulas makes the links of the polarization to the local and to the mean fields equivalent. However, in this way our Eqs. (3.51) and (3.52) become proved only for a discrete set of observation points—lattice nodes. Well, we have also discussed that we may spread all fields and polarizations to the whole volume of the array and this makes our equations correct in the bulk of the array. But near the array effective surface there is no guarantee that the relations between these spread functions keep the same

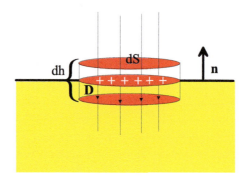

Figure 3.2 Illustration to the derivation of the bound charges through the p-dipole and quadrupole polarization densities.

as they were derived for the multipole centers. And they are really not applicable at the effective surface whose exact location we will define below. Let us show it, omitting (for simplicity of writing) the terms of the second order of smallness.

In free space and on the array effective boundary vectors **D** and **H** are not ambiguous and equal to: $\mathbf{D} = \epsilon_0 \mathbf{E}$, and $\mathbf{H} = (\mu_0)^{-1}\mathbf{B}$. We have to match them with their analogues under the boundary. Let us write relation (3.51) keeping the terms of the zero and first orders for the effective volume occupied by the array:

$$D_\alpha = \varepsilon_h E_\alpha + P_\alpha - \frac{1}{2}\nabla_\beta Q_{\alpha\beta}. \tag{3.55}$$

At the effective boundary of this volume there are spread bound charges with the surface density σ. It is commonly believed that σ is related to the p-dipole polarization as $\sigma = P_n$, where **n** is the external normal to the surface. However, it is so only for media with purely p-dipole polarization. Let us apply the Gauss theorem to a cylindrical elementary volume $dV = dS\, dh$, shown in Fig. 3.2 taking into account that the divergence of **D** is identically equal to zero. From (3.55) we have

$$\nabla \cdot \mathbf{D} = \varepsilon_h \nabla \cdot \left(\mathbf{E} + \frac{\mathbf{P}}{\varepsilon_h} - \frac{1}{2\varepsilon_h}\nabla \cdot \overline{\overline{Q}} \right) = 0. \tag{3.56}$$

Next, we substitute Maxwell's equation for the mean field

$$\nabla \cdot \mathbf{E} = -\frac{\rho}{\varepsilon_h},$$

where ρ is the mean density of the polarization charges, into (3.56) and obtain

$$dS \int_{-dh/2}^{dh/2} \rho \, dz = \left(\mathbf{P} - \frac{1}{2} \nabla \cdot \overline{\overline{Q}} \right)_n .$$

The left-hand side is by definition equal to the surface bound charge density σ, and in the right-hand side the p-dipole and quadrupole polarizations are taken on the bottom surface of the elemental cylinder (on its top $\mathbf{P} = 0$ and $\overline{\overline{Q}} = 0$). We have then

$$\sigma = P_n - \frac{1}{2} (\nabla \cdot \overline{\overline{Q}})_n. \qquad (3.57)$$

This formula points out the model inconsistency at the effective boundary of the medium. In fact, σ is a covariant, physically measurable value whereas the right-hand side of (3.57) is dependent on the selection of the multipole centers of particles located in the vicinity of the boundary. Recall that in our model the macroscopic multipole density results from the integration of the microscopic one, which depends on this choice. The model allows a shift of the centers of these particles by any vector $\mathbf{R}$ so that $R < d/2$ (so that the particles centers are still inside the corresponding unit cells). Then all microscopic multipoles of the medium will change, and the quadrupoles, too. Of course, if this change is negligibly small and the right-hand side of (3.57) will change only slightly it can be attributed to an admissible error of the homogenization model and we will be happy with our Eqs. (3.51) and (3.52).

However, it is not the case. Let us rewrite the relation (3.18) for the macroscopic polarization density at the old center $\mathbf{r} = 0$ of any unit cell:

$$Q_{\alpha\beta} = \frac{1}{j\omega V} \int_V (J_\alpha r_\beta + J_\beta r_\alpha) \, dV.$$

and substitute the relation

$$J_\alpha(\mathbf{r}) = E_\beta(\mathbf{r}) \int_\Omega K_{\alpha\beta}(\mathbf{r} - \mathbf{r}') \, dV' \qquad (3.58)$$

following from (3.22) in the first-order approximation. Then we obtain the quadrupole susceptibility of the medium in the following form:

$$Q'_{\alpha\beta\gamma} = \frac{1}{j\omega V} \int_V \int_\Omega (K_{\alpha\beta} r'_\gamma + K_{\beta\gamma} r'_\alpha) \, dV' dV. \qquad (3.59)$$

Now let us assume that the multipole center of the reference particle is shifted from $\mathbf{r} = 0$ to point $\mathbf{r} = \mathbf{R}$. Then the quadrupole susceptibility of the medium becomes

$$Q'_{\alpha\beta\gamma} = \frac{1}{j\omega V} \int_\Omega \int_{V_p} \left(K_{\alpha\beta} (r'_\gamma + R_\gamma) + K_{\beta\gamma} (r'_\alpha + R_\alpha) \right) dV' dV$$

$$= \frac{1}{j\omega V} \int_\Omega \int_{V_p} \left(K_{\alpha\beta} r'_\gamma + K_{\beta\gamma} r'_\alpha \right) dV' dV$$

$$+ \frac{R_\gamma}{j\omega V} \int_\Omega \int_{V_p} K_{\alpha\beta} \, dV' dV + \frac{R_\alpha}{j\omega V} \int_\Omega \int_{V_p} K_{\beta\gamma} \, dV' dV.$$

(3.60)

The first integral term in the right-hand side of (3.60) is the old quadrupole susceptibility (3.59). Two last integral terms in the right-hand side of (3.60) is its change due to the center shift. These terms can be expressed through the p-dipole susceptibility. In fact, according to (3.17), we have

$$P_\alpha = \frac{1}{j\omega V} \int_\Omega J_\alpha \, dV,$$

and in the zero order of smallness it means that

$$P_\alpha = \frac{E_\alpha(0)}{j\omega V} \int_\Omega \int_{V_p} K_{\alpha\beta} \, dV' dV.$$

Therefore, for the zero-order dipole susceptibility we may write

$$a_{\alpha\beta} = \frac{1}{j\omega V} \int_\Omega \int_{V_p} K_{\alpha\beta} \, dV' dV.$$

This allows us to rewrite (3.60) is the form

$$(Q'_{\alpha\beta\gamma})^{\text{new}} = (Q'_{\alpha\beta\gamma})^{\text{old}} + \Delta Q'_{\alpha\beta\gamma}, \qquad (3.61)$$

where it is denoted

$$\Delta Q'_{\alpha\beta\gamma} = R_\alpha a_{\beta\gamma} + R_\gamma a_{\alpha\beta}, \qquad (3.62)$$

and we see that the change in the quadrupole susceptibility $\Delta Q'_{\alpha\beta\gamma}$ is the value of the first order of smallness. In fact, $|\mathbf{R}|$ is the value of the order of d, which is a first-order value in our theory, and the p-dipole susceptibility a_{ij} is a zero-order value. So, $(Q'_{\alpha\beta\gamma})^{\text{old}}$ and $\Delta Q'_{\alpha\beta\gamma}$ have the same order of magnitudes. This conclusion keeps valid if we include into our analysis the terms of the second order of smallness we have omitted for simplicity. The change in the surface charge due density turns out to be not negligible since in the theory which pretends to the account of the second-order terms, we cannot neglect the first-order terms. This means that the definitions of the auxiliary electromagnetic vectors are not consistent with the purposes of the theory. If we want to apply for finite-thickness arrays our auxiliary electromagnetic vectors must be redefined.

Notice that the m-dipole susceptibility is also sensitive to the arbitrary choice of the particle effective center. Similarly, one can show that the shift of this center by a vector **R** results in the change of the tensor $\bar{\bar{G}}$ by the following additive value:

$$\Delta G_{ij} = \frac{j\omega}{2} e_{jkm} R_k a_{im}. \tag{3.63}$$

Also, it is worth to notice that from (3.62) it follows the change of the tensor $\bar{\bar{a}}^{\text{sym}}$, entering (3.48). Non-covariant equations (3.51) and (3.52) were named in [1] *quasi-material equations*.

To conclude this subsection we have to make two important remarks. First, the full expansion (3.25) as a whole is covariant. The non-physical equation (3.57) resulted from the truncation of the infinite expansion. Unfortunately, this truncation is the same as the analytical model of WSD. Therefore, the redefinition of the auxiliary field vectors resulting in covariant MEs is needed.

Our second remark is as follows. The non-covariance in the first and second orders of smallness with respect to d/λ arises only if the microscopic polarization currents in the particles are not closed loops. There are situations when these currents are practically closed loops. For example, in properly designed canonical CPs and OPs the loops are only slightly open because the gap between the dipole arms is very small. Numerical studies have shown that both electric and magnetic quadrupoles of properly designed CPs and OPs are negligible compared to the p- and m-dipoles in the frequency region of the dipole resonances [79]. The same observation refers to the ceramic spheres. In media with the p-m dipole response (BA or non-BA) the non-covariance in (3.51) and (3.52) does not arise.

However, if the artificial magnetism and/or bianisotropy arise in a loop with large open part or in an asymmetric pair of metal strips shown in Fig. 2.3e, the situation charges drastically. Let the strips of this dimer be stretched along the z-axis. For such a pair the magnetic moment directed along y is exactly equal to the xz-component of the electric quadrupole moment. It is clear from definitions (3.18) and (3.20) (see also our comment above on the dimensionless normalization of multipoles). Therefore, a composite medium formed by such dimers cannot be described as a simple Ω-medium with only the p-m-dipole response. Quadrupoles are obviously present in the MEs of such a medium except, perhaps, a specific frequency at which its p-dipole polarization vanishes. Below we again discuss the composite media of dimers because they are very important MMs—if properly designed they offer magnetism in the visible range.

3.4.3 Material Equations Covariant in the First Order of WSD

In order to make **D** and **H** covariant at least within the first-order approximation of WSD we should add to **D** and **H** defined by relations (3.10), (3.11), (3.12) and (3.13), in other words by expansions (3.40) and (3.41) two certain vectors, denoted below as **K** and **T**, respectively. These vectors should be chosen so that the coordinate dependence of M'_{ij} and $(Q'_{kij} - Q'_{kji})$ in Raab's equations (3.51) and (3.52) is compensated. As it was already explained, Maxwell's equations will be not violated with this redefinition of **D** and **H** if these two additional vectors are related as

$$\mathbf{K} = \frac{1}{j\omega} \nabla \times \mathbf{T}.$$

The needed vectors **K** and **T** can be guessed:

$$K_i = \frac{1}{2}(Q'_{jik} - Q'_{ijk})\nabla_j E_k, \quad T_i = -\frac{j\omega}{2} e_{ijk} Q'_{jkm} E_m. \tag{3.64}$$

The operation

$$\mathbf{D}^{\text{new}} = \mathbf{D}^{\text{old}} + \mathbf{K}, \quad \mathbf{H}^{\text{new}} = \mathbf{H}^{\text{old}} + \mathbf{T} \tag{3.65}$$

applied to (3.51), (3.52) leads to the following equations (terms of the second order are not shown):

$$D_i = \epsilon_{ij} E_j + \frac{1}{2} e_{ijk} e_{klm} Q'_{mls} \nabla_j E_s - g_{ij} B_j + \cdots \tag{3.66}$$

and
$$H_i = \mu_0^{-1} B_i - (M'_{ij} + \frac{j\omega}{2} e_{jkm} Q'_{mik}) E_j + \cdots \quad (3.67)$$

Using relation (3.50), Eqs. (3.66) and (3.67) can be rewritten in the form

$$D_i = \epsilon_{ij} E_j - [g_{ij} + \frac{j\omega}{2} e_{ikm} Q'_{mjk}] B_j + \cdots \equiv \epsilon_{ij} E_j + j\xi_{ij} B_j, \quad (3.68)$$

$$B_i = \mu_0 H_i + \mu_0 \left(M'_{ij} + \frac{j\omega}{2} e_{jkm} Q'_{mik} \right) E_j \equiv \mu_0 H_j - j\mu_0 \xi_{ji} E_j. \quad (3.69)$$

The tensor form of these MEs is more compact:

$$\mathbf{D} = \bar{\bar{\epsilon}} \cdot \mathbf{E} + j\bar{\bar{\xi}} \cdot \mathbf{B} + \cdots, \quad (3.70)$$

$$\mathbf{B} = \mu_0 \mathbf{H} - j\mu_0 \bar{\bar{\xi}}^T \cdot \mathbf{E} + \cdots. \quad (3.71)$$

Here in order to get $\bar{\bar{\xi}}^T$ in (3.69) and (3.71) we have imposed the condition

$$M'_{ij} = -g_{ji}. \quad (3.72)$$

Here we temporary postulate it noticing that otherwise, we cannot implement the reciprocity in our MEs (see above). However, below we will prove that M'_{ij} is really equal to $-g_{ji}$ deriving the so-called generalized Maxwell Garnett mixing formulas.

In Eqs. (3.70) and (3.71) ξ_{ij} denotes $jg_{ij} - \omega e_{ikm} Q'_{mjk}/2$ that can be also presented as

$$\xi_{ij} = -jM'_{ji} - \frac{\omega e_{ikm} Q'_{mjk}}{2}. \quad (3.73)$$

As everywhere $\bar{\bar{\xi}}^T$ in (3.71) denotes the transposed $\bar{\bar{\xi}}$. In lossless media tensor $\bar{\bar{\xi}}$ is purely real (see above), and this is why we introduced the imaginary unity j in the definition of this tensor by Eqs. (3.68) and (3.69). In isotropic media $\bar{\bar{\xi}} = \xi \bar{\bar{I}}$, i.e., MEC parameter is the scalar parameter ξ. In this case it is called the *chirality* parameter (see above).

One can see that the first-order spatial dispersion still does not give magnetic susceptibility of composite or molecular media. The material parameter $\bar{\bar{\xi}}$ is the magneto-electric coupling parameter. Since we consider reciprocal media, the same MEC parameter $\bar{\bar{\xi}}$

enters both material equations (3.70) and (3.71). It contains the magneto-dipole susceptibility to the uniform (across the unit cell) part of the averaged electric field, which is equal to the electro-dipole susceptibility to the vortex part of the averaged electric field.[c] It also contains the quadrupole susceptibility to the uniform part of the electric field.

Using Maxwell's equations (3.31) and (3.32) Eqs. (3.70), (3.71) after some easy algebra can be rewritten as

$$\mathbf{D} = \overline{\overline{\varepsilon}}' \cdot \mathbf{E} + j\overline{\overline{\chi}} \cdot \mathbf{H}, \quad \mathbf{B} = \overline{\overline{\mu}} \cdot \mathbf{H} - j\overline{\overline{\chi}} \cdot \mathbf{E}. \quad (3.74)$$

This is nothing but the Lindell–Sihvola MEs for media BA without artificial magnetism. This formalism is very popular in the literature, however, it is physically irrelevant. In (3.74) the MEC parameter $\overline{\overline{\chi}}$ includes not only first-order parameters $\overline{\overline{M}}'$ and $\overline{\overline{Q}}'$ as our MEC parameter $\overline{\overline{\xi}}$, but also the p-dipole susceptibility $\overline{\overline{a}}$ which is the zero-order parameter of smallness. Moreover, the nontrivial permeability $\overline{\overline{\mu}}$ in (3.74) appears and comprises the p-dipole susceptibility $\overline{\overline{a}}$ (zero-order parameter) and the electric quadrupole susceptibility $\overline{\overline{Q}}'$ (first-order parameter). However, we already know that the artificial magnetism is the effect of the second order. Therefore, the most consistent form of ME for media with first order of SD is namely the system (3.70), (3.71). Below we will see that it is the special case of the so-called Post formalism—Post's equations for BA media without artificial magnetism.

3.4.4 Material Equations Covariant in the Second Order of WSD

Now let us rewrite Eqs. (3.70) and (3.71) including in them the second-order terms from Eqs. (3.51) and (3.52) that the operation (3.65) keeps intact:

$$D_i = \varepsilon_{ij} E_j + j\xi_{ij} B_j + \beta_{ijkl} \nabla_j \nabla_k E_l \quad (3.75)$$

and

$$H_i = \mu_0^{-1} B_i + j\xi_{ji} E_j + \gamma_{ijk} \nabla_j E_k. \quad (3.76)$$

---

[c]i.e., to the uniform across the unit cell part of the magnetic field.

where tetradic tensor with components β_{ijkl} and triadic tensor γ_{ijk} are defined through corresponding multipole susceptibilities of the medium by (3.53). First, in the same way as we did above let us share out the antisymmetric part from the tensor $\overline{\overline{\overline{M}}}''$:

$$M''_{ijk} = (M''_{ijk})^{\text{symm.}}_{ijk} + (M''_{ijk})^{\text{asym.}}_{ijk} \equiv f_{ijk} + \frac{2}{j\omega} e_{ikm} G_{jm}, \quad (3.77)$$

where again the antisymmetric tensor has been presented through the Levi–Civita triadic and a certain dyadic $\overline{\overline{G}}$ whose properties are not yet known. Above we have used the relations (3.15) for the symmetric part of the electro-dipole susceptibility to the mean field derivatives and the electro-quadrupole susceptibility to the mean field. Below we present the detailed derivation of this important relation.

Similar relations were derived for the symmetric part of the magneto-dipole susceptibility to the mean field derivatives and the magneto-quadrupole susceptibility to the mean field:

$$f_{ijk} = f_{jik} = S'_{jik} = S'_{jki}. \quad (3.78)$$

Microscopic analogues of relations (3.78) were obtained in work [165]. Deduced from those microscopic formulas and CMLL formulas macroscopic relations (3.78) were derived in [93]. This important result will be discussed below together with the generalized Maxwell Garnett mixing rules.

Equation (3.76) can be rewritten after substitutions of (3.77) and (3.78) in the form:

$$H_i = (\mu_0^{-1} I_{ij} + G_{ij}) B_j + j\xi_{ji} E_j + (S'_{jik} - S'_{ijk}) \nabla_j E_k. \quad (3.79)$$

Of course in (3.79) we again took into account the relation (3.50).

Now we can already see the reason of the non-trivial permeability: it comes from the anti-symmetric part of the susceptibility of the magnetic dipole moment to spatial derivatives of the electric field. This is the susceptibility of the m-dipole moment to the curl of **E**, i.e., to **B**. The susceptibility of the magnetic quadrupole moment to **E** is symmetric with respect to two last indexes as well as that of the electric quadrupole moment.

Equations (3.75) and (3.79) are still not MEs since tensors with components G_{ij} and $(S'_{jik} - S'_{ijk})$ are sill origin-dependent (not covariant in the second order). It is possible to show (similarly as it was done above in the first-order theory) that these constitutive equations will result in the error of the second order of smallness for the bound charges on the medium surface. Therefore, we have to find a vector $\mathbf{T}'$ so that the operation

$$\mathbf{D}^{\text{new}} = \mathbf{D}^{\text{old}} + \nabla \times \mathbf{T}', \quad \mathbf{H}^{\text{new}} = \mathbf{H}^{\text{old}} + j\omega \mathbf{T}' \tag{3.80}$$

would give new $\mathbf{D}$ and $\mathbf{B}$ which are covariant in the second order of smallness.

At this point we can return to the expansion (3.24) and notice that all the coefficients in this series must be origin-independent. This is because the current $\mathbf{J}$ and the mean field and its derivatives in (3.24) are all measurable quantities [92] (unlike the multipole moments). The comparison of (3.24) and (3.25) gives a set of equations relating covariant coefficients b_{ij}, b_{ijk}, b_{ijkl} with non-covariant multipole susceptibilities:

$$b_{ij} = a_{ij}, \quad b_{ijk} = \frac{1}{2}(a'_{ijk} - Q'_{ikj}) + \frac{1}{j\omega} e_{ikn} M'_{nj}, \tag{3.81}$$

$$b_{ijkl} = \frac{1}{6}(a''_{ijkl} - Q''_{ikjl} + O'_{iklj}) + \frac{1}{2j\omega} e_{ikn}(M''_{njl} - S'_{nlj}). \tag{3.82}$$

The last equation rewritten in the form

$$b_{ijkl} = \frac{1}{6}(a''_{ijkl} - Q''_{ikjl} + O'_{iklj}) + \frac{1}{2j\omega} e_{ikn}(S'_{jln} - S'_{nlj}). \tag{3.83}$$

is the covariant combination of the multipole susceptibilities that we need. We can notice that adding the term

$$K'_i = (\nabla \times \mathbf{T}')_i = \frac{1}{2j\omega} e_{ikn}(S'_{jln} - S'_{nlj}) \nabla_k E_j \tag{3.84}$$

to $\mathbf{D}^{\text{old}}$ defined by Eq. (3.75) we obtain $\mathbf{D}^{\text{new}}$ in the form

$$D_i^{\text{new}} = \varepsilon_{ij} E_j + j\xi_{ij} B_j + b_{ijkl} \nabla_l \nabla_k E_j. \tag{3.85}$$

This equation is ME because it contains the terms of the second order with covariant coefficients.

It is possible to show (see [93]) that the operation (3.80) with vector $\mathbf{T}'$ that can be found from Eq. (3.84) removes from Eq. (3.79) the term $(S'_{jik} - S'_{ijk}) \nabla_j E_k$ and simultaneously adds to G_{ij} a dyad that makes the corresponding coefficient origin-independent. The result of the redefinition (3.80) with substitution of (3.84) gives two MEs.

The first one simply reproduces (3.85) and the second one is that for **H**:

$$D_i = \varepsilon_{ij} E_j + j\xi_{ij} B_j + b_{ijkl} \nabla_l \nabla_k E_j, \qquad (3.86)$$

$$H_i = (\mu_{ij})^{-1} B_j + j\xi_{ji} E_j. \qquad (3.87)$$

These relations are MEs of media with the second-order SD. Here the following notation has been introduced:

$$(\mu^{-1})_{ij} = \frac{1}{\mu_0} I_{ij} + G_{ij} - \frac{j\omega}{2} e_{jkm} S'_{imk}. \qquad (3.88)$$

It is clear that the inverse tensor to $(\overline{\overline{\mu}}^{-1})$ is the absolute permeability of the effective medium in which the magnetic susceptibility arises as an effect of the second-order spatial dispersion. Since we suppose the multipole hierarchy, the terms of the second order are small compared to the terms of the zero order, and in (3.88)

$$|G_{ij} - \frac{j\omega}{2} e_{jkm} S'_{imk}| \ll 1.$$

Next, assume that the imaginary part of the tensor $\overline{\overline{G}}$ and the real part of the tensor $\overline{\overline{S}}'$ are small enough.[d] Then there is a coordinate frame, in which tensor $\overline{\overline{\mu}}$ can be present in a diagonal form ($\mu_{ij} = 0$ if $i \neq j$) and formula (3.88) can be approximated as

$$\mu_{ii} \approx \mu_0 \left(1 - \mu_0 G_{ii} + \frac{j\omega\mu_0}{2} e_{ikm} S'_{imk}\right). \qquad (3.89)$$

One can guess here that the dyadic tensor $\mu_0 \overline{\overline{G}}$ taken with sign minus is the relative susceptibility of the magnetic dipole polarization to the vortex part of the mean electric field, i.e., (effectively) to the magnetic field. As to the triad $\overline{\overline{S}}$, it is the susceptibility of the magnetic quadrupole polarization *to the uniform part of the mean electric field*. Thus, not only the magnetic susceptibility, also the magnetic multipole response to the electric field contributes into the effective permeability of multipolar media.

There is nothing physically inconsistent. The theory of WSD presents something which is in fact the response to the electric

---

[d]This assumption as it follows from the generalized Maxwell Garnett model corresponds to low magnetic losses in the effective medium.

field of the wave in a form of the effective response to the magnetic field. The goal of this theory is to describe media with SD as effectively local media, and this goal is achieved. The requirement that all parameters in Eqs. (3.86) and (3.87) are covariant is crucial for this goal, whereas the desire of a reader to have the effective permeability fully determined by the microscopic magnetic response is not. This desire will be satisfied only in absence of magnetic quadrupoles.

The term $b_{ijkl}\nabla_l\nabla_k E_j$ does not comprise spatial derivatives of the magnetic field, i.e., terms $\nabla\nabla\times\mathbf{E}$. In fact, the tetrad b_{ijkl} is invariant to the commutation of indices (j, k) and (j, l). It follows from its multipole decomposition (3.83) and from the reciprocity of the multipolar susceptibilities (proved below). In other words, our theory expressed by Eqs. (3.86) and (3.87) does not allow the bianisotropy in the second order of SD. Bianisotropy is purely the first-order effect.

3.4.5 Some Special Cases of MEs in Media with WSD

When the medium contains only p-dipole and m-dipole scatterers EMPs entering Eqs. (3.86) and (3.87) simplify. First, we have $b_{ijkl} = 0$. Second, the MEC parameter

$$\xi_{ij} = jM_{ji} - \omega e_{ikm}Q'_{mjk}$$

simplifies to $\overline{\overline{\xi}} = j(\overline{\overline{M}})^T$ and the absolute permeability (3.89) becomes as simple as $\overline{\overline{\mu}} = \mu_0(\overline{\overline{I}} - \mu_0\overline{\overline{G}})^{-1}$. In this case Eqs. (3.86) and (3.87) take the form:

$$\mathbf{D} = \overline{\overline{\varepsilon}}\cdot\mathbf{E} + j\overline{\overline{\xi}}\cdot\mathbf{B}, \quad \mathbf{B} = \overline{\overline{\mu}}\cdot(\mathbf{H} - j\overline{\overline{\xi}}\cdot\mathbf{E}), \qquad (3.90)$$

called the *Post equations* for BA media. These equations were, for the first time, phenomenologically suggested in the classical book [41]. Here, we have derived them from the concept of WSD. Notice that (3.90) refers to BA media with artificial magnetism. In absence of artificial magnetism, they describe the media with SD of the first order and become Eqs. (3.70) and (3.71).

If the scatterers possess no BA response, the MEC parameter vanishes and we come to MEs of a magneto-dielectric

$$\mathbf{D} = \overline{\overline{\varepsilon}}\cdot\mathbf{E}, \quad \mathbf{B} = \overline{\overline{\mu}}\cdot\mathbf{H}. \qquad (3.91)$$

The necessary condition of the validity of (3.90) is the absence of higher multipoles. Only for media formed by p- and m-dipoles the theory of WSD results in the Post equations.

In the isotropic media with SD of the second order, there can be higher multipoles as well. From the symmetry, an only possible isotropic representation of the term $b_{ijkl}\nabla_l\nabla_k E_j$ in Eq. (3.86) is $b\,\mathrm{grad}\nabla\cdot\mathbf{E}$, which gives:

$$\mathbf{D} = \varepsilon\mathbf{E} + j\xi\mathbf{B} + b\nabla\nabla\cdot\mathbf{E}, \qquad (3.92)$$

$$\mathbf{H} = \mu^{-1}\mathbf{B} + j\xi\mathbf{E}. \qquad (3.93)$$

Substituting the second ME into the first one we can rewrite these equations in the form generalizing the Lindell–Sihvola MEs for chiral media:

$$\mathbf{D} = \varepsilon'\mathbf{E} + j\xi'\mathbf{H} + b\nabla\nabla\cdot\mathbf{E}, \qquad (3.94)$$

$$\mathbf{B} = \mu\mathbf{H} - j\xi'\mathbf{E}, \qquad (3.95)$$

where Lindell–Sihvola's EMPs are expressed through EMPs of our molecular theory as follows:

$$\varepsilon' = \varepsilon + \xi^2\mu, \qquad \xi' = \xi\mu. \qquad (3.96)$$

It is clear that the Lindell–Sihvola constitutive formalism is physically relevant. Although it is most popular in the literature, the corresponding EMPs do not allow one to understand that the electric susceptibility is the zero-order effect and that the chirality is the first-order effect. In this formalism the all effects are mixed up.

In fact, a chiral medium without artificial magnetism is a reciprocal isotropic medium with SD of the first order. Chiral medium with artificial magnetism manifests WSD of both first and second orders. Besides the artificial magnetism the second-order effect of SD in isotropic media can be also manifested by a new parameter b. This b is a valid EMP because the term $b\nabla\nabla\cdot\mathbf{E}$ cannot be removed from (3.94) by further redefinition of $\mathbf{D}$ and $\mathbf{H}$. Whatever redefinition removing this terms from $\mathbf{D}$ one would try it will obviously violate the covariance of MEs.

The same observations can be done for the anisotropic case described by MEs (3.86) and (3.87). In principle, these equations can be also expressed in the generalized Lindell–Sihvola form:

$$D_i = \varepsilon'_{ij}E_j + j\xi'_{ij}H_j + b_{ijkl}\nabla_k\nabla_l E_j, \qquad (3.97)$$

$$B_i = \mu'_{ij} H_j - j\xi'_{ji} \mathbf{E}. \tag{3.98}$$

Again, in the presence of higher multipoles each of three tensors—$\overline{\overline{\varepsilon}}'$, $\overline{\overline{\mu}}'$ and $\overline{\overline{\xi}}'$—comprises the terms of the zero, first and second orders of smallness mixed up. The same drawback holds for the Drude-Born-Fedorov formalism. Only the generalized Post formalism expressed by our Eqs. (3.86) and (3.87) is free of this mishmash. Our $\overline{\overline{\varepsilon}}$ is the zero-order parameter, our $\overline{\overline{\xi}}$ is the first-order parameter, our $\overline{\overline{\mu}}$ and $\overline{\overline{\overline{b}}}$ are the second-order parameters.

3.4.6 Misinterpretation of Multipolar Composites Treated as Magneto-Dielectric Media

This subsection is absent in the original book [1] and is the author's response to the publications on MMs which contain another common mistake—identification of the multipole response with artificial magnetism.

If a constitutive particle is a sphere or a canonical CP, OP or even such an SRR in which the induced polarization current is almost uniform in azimuthal representation practically (this refers, for example, to SRRs studied in our work [170]) the particle response can be represented by a pair of a p-dipole and a m-dipole. Such a media can be really described (in the frequency region where the SD is weak) by Eq. (3.90). In the special case when the bianisotropy is compensated by the special arrangement of constitutive particles the MEs take the form (3.91). And a mixture with random orientations of particles is an isotropic magneto-dielectric.

However, if the constitutive particle is S-shaped as in [167], is a single split ring with large open part (C-shaped particle) or a U-shaped particle as in works [175–177], the polarization current in it corresponds to significant bound charges induced at the sides of a split. Then the polarization current cannot be modeled as a uniform azimuthal current flowing in an effective loop. Therefore, the response of such SRRs is essentially multipolar. To share the non-electric response from the electromagnetic one is possible, but it is erroneous to treat it as the magnetic one like it was done in [175, 176]. If we assume that SD in the medium of such particles is weak the magnetic response of the medium should be shared between parameters $\overline{\overline{\mu}}$ from (3.87) (permeability) and $\overline{\overline{b}}$ from (3.86). The last parameter has no commonly adopted name yet, and can be named, for example, *b-parameter*. The magnetic response in the case of nonzero b-parameter obviously imply the magnetic quadrupoles induced by the electric field. The corresponding susceptibility tensor enters both b-parameter and permeability. As to the MEC parameter ξ, if we consider the effective medium suggested in [167], this EMP is zero.

The impact of higher multipoles is especially spectacular for effective media formed by dimers of resonant electric scatterers. They can be pairs of closely positioned plasmonic nano-pyramids, nano-wires and nano-plates (see [171–173]). In this case the consolidated contribution of higher multipoles dominates over the magnetic dipole response. In this case, the use of Eq. (3.91) instead of Eqs. (3.86) and (3.87) leads to even more weird frequency dependence of the retrieved material parameters than that manifested by the antiresonance.

Let us consider two examples of a complex "magnetic" particle for the microwave and visible ranges of frequencies, both realized as a closely positioned pair of small scatterers. In the first case (see Fig. 3.3a,b) these are metal strips located so that they are coupled capacitively. In the second case (see Fig. 3.3a,b) these scatterers are plasmonic nanospheres. As it was shown in [171] and in precedent works, within the band of the so-called *plasmonic resonance* of the individual nanoparticle there is a frequency ω_m at which the magnetic mode is excited in the pair. This mode corresponds to the antiparallel excitation of resonant electric dipoles in two nanoparticles. As a result, the total p-dipole moment of the nanopair at this mode is zero, and the particle can be presented as a superposition of a m-dipole and an electric quadrupole, both with susceptibilities to local field $\mathbf{E}^{loc}$ and its spatial derivatives, and a magnetic quadrupole and electric octopole, both with susceptibilities to $\mathbf{E}^{loc}$.[e]

---

[e]Here we assumed by default that the sizes of the particles and the distance between them are so that the electric octopole moment can be neglected, as well as the electric quadrupole polarization by the magnetic field and the magnetic quadrupole polarization by the electric one.

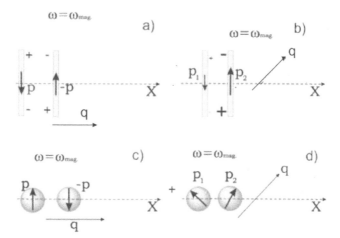

Figure 3.3 A pair of two scatterers impinged by a wave propagating with wave vector q at frequency $\omega = \omega_{mag}$. Subfigures (a) and (b) correspond to the dual-wire particle. Subfigures (c) and (d)—to a plasmonic dimer. When the wave propagates along x a purely magnetic mode is excited. The off-diagonal quadrupole moment and the z-oriented m-dipole moment are equivalent one another (left panel). When the wave propagates obliquely (right panel) the superposition of the p-dipole, m-dipole and quadrupole modes is excited.

Let the dimer be excited by a wave propagating along the axis x, as it is shown in Fig. 3.3a,c. Then at a certain frequency $\omega = \omega_{mag}$ the phase shift of the wave between two p-dipole scatterers correspond to the condition $p_2 = -p_1$, and the total electric dipole moment of the dimer vanishes. In this case both the magnetic moment directed along y (along the magnetic field vector of the wave) and the electric quadrupole moment of the nanopair are origin-independent. It follows from our formulas (3.62) and (3.63) in which the dipole susceptibility tensor $\bar{\bar{a}}$ vanishes. In other words, the zero polarizability of the p-dipole leads to the independence of **m** on the location of the particle center. Of course the component Q_{xz} of the quadrupole tensor is nonzero and (being normalized as it was discussed above) equals to m_y. However, its contribution in our MEs at this frequency vanishes when the magnetic polarizability is covariant! The dimer excitation holds only due to the nonzero curl of $\mathbf{E}^{loc}$ at the dimer center that can be interpreted as the excitation by local magnetic field (artificial magnetism). Therefore, the medium *for this special case of propagation and at this specific frequency* can be described as the medium with artificial magnetism. For the waves propagating in the directions orthogonal to the x-axis the excitation of both metal elements is in-phase, it is a dielectric medium.

However, it would be a serious mistake to believe that the medium of dimers is simply a resonant magneto-dielectric as it was believed in [171] and [172]. First, beyond the special frequency ω_{mag} the total dipole moment will be nonzero even for the wave traveling along x. And both m-dipole and electric quadrupole moments become non-covariant. Second, if the propagation is oblique as in Fig. 3.3b,d, the phase shift of the wave over our dimer even at ω_{mag} will be different from that corresponding to the propagation along x. Therefore, the response for the oblique case will be not antisymmetric and $p_2 \neq -p_1$. Again, the nonzero p-dipole excited in the dimer makes MEs of our medium non-covariant if we define the auxiliary field vectors in a traditional way and represent the effective medium as a magneto-dielectric material. We have to use Eqs. (3.86) and (3.87) with the terms $\nabla\nabla \cdot \mathbf{E}$ in all cases if $\omega \neq \omega_{\text{mag}}$ and even if $\omega = \omega_{\text{mag}}$ if we consider the waves propagating not only along x.

What if we express the dispersion characteristics of such media in terms of $\overline{\overline{\varepsilon}}$ and $\overline{\overline{\mu}}$ neglecting the b-parameter in (3.86)? Then we may drastically exaggerate the artificial "magnetism" attributing it to the frequencies at which it is absent The same remark concerns the magnetism of optical SRRs reported in works [175–178] and other similar works. The use of Eq. (3.91) instead of Eqs. (3.86) and (3.87) resulted in weird frequency dependencies of EMPs that were obtained in [171, 172, 175–178] fitting experimental results to the wrong model of a magneto-dielectric medium. And the effective permeability in these works may take negative values or attain local maxima at frequencies where its correct value is close to unity (see on this in [174]).

However, our criticism mainly refers to the dimers with the overlapping resonances of the p-dipole and m-dipole responses. There are some plasmonic dimers in which the resonance bands of the m-dipole and p-dipole modes are well separated on the frequency axis [173]. Then the magnetic mode so strongly dominates at the magnetic resonance that the parasitic excitation of the p-mode is negligible in the whole resonance band. For such an MM, the band of the resonant permeability corresponds to the dispersion-free permittivity (the MM is a resonant artificial magnetic). And the band of the resonant permeability corresponds to the permeability equal to unity (the MM is a resonant artificial dielectric).

Also, optical SRRs with four narrow splits of a plasmonic nanoring (such SRRs were suggested in [179]) are practically free of higher-order multipoles. These splits are highly capacitive, the displacement currents are strongly concentrated inside the splits and efficiently close the effective current loop making its polarization current uniform. No significant charges in this case are accumulated at the splits edges. In this situation the quadrupole and other multipoles should be negligible and it is worth noticing that the frequency dependence of the effective permeability obtained as the fitting parameter in [179] is physically sound.

Finally, it is worth to notice that in some papers the permeability—scalar as in [180] or tensor as in [181]—is attributed to photonic crystals in order to describe their SD in the vicinity of the Bragg mode. Namely, the Bragg passband is explained in terms of the positive permittivity and negative permeability. Of course such continuous media are also opaque but the reason of the opacity is different. We will postpone the discussion of this point to the last chapter of our book where we will point out the difference between an effectively continuous medium opaque at the resonance of its inclusions and a self-resonant photonic crystal opaque due to the Bragg reflective resonance.

3.5 On Two Equivalent Approaches to WSD

Now let us discuss how our theory corresponds to the commonly adopted insight of SD as the dependence of EMPs on the eigenmode wave vector $\mathbf{q}$. In space of wave vectors the spatial derivative is replaced by multiplication by $-j\mathbf{q}$ (scalar product for divergence and vector one for curl). For example, Raab's equation (3.51) can be rewritten for complex amplitudes of a spatial harmonic of the electromagnetic field (for spatial harmonics we use a tilde in our notations):

$$\tilde{D}_i(\mathbf{q}) = \epsilon_0 \tilde{E}_i(\mathbf{q}) + a_{ij}\tilde{E}_j(\mathbf{q}) + \frac{j}{2}((a_{ijm} - Q'_{mij})q_j\tilde{E}_m(\mathbf{q})$$
$$+ \frac{1}{6}(a''_{iljm} - Q''_{ijml} - O'_{imlj})q_j q_m \tilde{E}_l(\mathbf{q}) + \cdots$$

The same relation is more compact in the tensor form:

$$\tilde{\mathbf{D}} = \overline{\overline{\varepsilon}} \cdot \tilde{\mathbf{E}} + (\overline{\overline{\overline{A}}} \cdot \mathbf{q}) \cdot \tilde{\mathbf{E}} + (\overline{\overline{\overline{\overline{B}}}} \cdot \mathbf{q} \cdot \mathbf{q}) \cdot \tilde{\mathbf{E}} \ldots, \qquad (3.99)$$

where the triad $\overline{\overline{\overline{A}}}$ and the tetrad $\overline{\overline{\overline{\overline{B}}}}$ have following components

$$A_{ijm} = \frac{j}{2}((a_{ijm} - Q'_{mij}), \qquad (3.100)$$

$$B_{ijlm} = \frac{1}{6}(a''_{iljm} - Q''_{ijml} - O'_{imlj}). \qquad (3.101)$$

Equation (3.99) is known in the theory of continuous media [5, 41, 53, 55] where it is treated as a formal definition of the spatial dispersion of the second order of smallness. In this theory in absence of the natural magnetism it is assumed that $\mathbf{H} = (\mu_0)^{-1}\mathbf{B}$, and the SD is by definition the dependence of the formally introduced effective permittivity on the argument $-q$. Then it is sometimes said (see [5]) that in practice the expansion (3.99) either diverges, and in this case the SD is strong and the medium is not effectively continuous, or converges, and then one may neglect all the terms of the order q^N with $N > 2$. This is the WSD of the second order (see also in [75]). This is, of course, correct, in this theory there is an only drawback—it is not very clear what physically means the field vector $\mathbf{D}$. And therefore, the physical meaning of the non-local permittivity in these classical books remains rather vague. To fill this theory with a clear physical content is simple involving our model. It is enough to see that tensor $\overline{\overline{\overline{A}}}$ is a combination of the medium susceptibilities given by formula (3.100), and tensor $\overline{\overline{\overline{\overline{B}}}}$ is expressed through the medium susceptibilities in accordance with (3.101).

Medium with SD of the 0-th order—medium without SD—is a dielectric. In this trivial case only one susceptibility $\overline{\overline{a}}$ is taken into account together with the permittivity and permeability of the ambient (ε_h and μ_0, respectively). This EMP $\bar{\bar{a}}$ is in our terminology the p-dipole susceptibility to the uniform mean field.

Medium with SD of the first order—BA medium—is characterized also by the triad $\bar{\bar{\bar{A}}}$ combining the p-dipole susceptibility to the solenoidal mean field and quadrupole susceptibility to the uniform mean field.

Medium with SD of the second order—is characterized by the permeability combining the m-dipole susceptibility to the solenoidal mean field and magnetic quadrupole susceptibility to the uniform mean field. However, it may be also characterized by tetrad $\bar{\bar{\bar{\bar{B}}}}$, combining three multipole responses and resulting in the mean field derivatives in our MEs.

All these results can be reformulated in terms of the **q**-dependence of the non-local permittivity. Notice that the formal approach based on the definition of auxiliary field vectors **H** and **D** by equations

$$\mathbf{D} = \varepsilon_h \mathbf{E} + \mathbf{J}, \quad \mathbf{H} = \mu_0^{-1} \mathbf{B} \qquad (3.102)$$

and spatial Fourier transform turned out to be fruitful for semiconductor crystals [75]. We have already mentioned that in these natural media the WSD of the second order is very different from the artificial magnetism. Practically, it results from the rather substantial size of the medium unit cell and is manifested in the excitation of polaritons at the surface and in the link between polaritons and excitons [75]. In this theory one assumed that the plane wave with wave vector **q** propagates in the medium, for which the expansion (3.24) for the mean polarization current can be written in the form

$$J_i = j\omega(b_{ij} E_j + j b_{ijk} q_k E_j - b_{ijkl} q_l q_k E_j + \cdots) \qquad (3.103)$$

Then substituting relation (3.103) into (3.102) we immediately obtain for the mean field vectors:

$$D_i = \varepsilon_{ij}(\omega, \mathbf{q}) E_j,$$
$$\varepsilon_{ij}(\omega, \mathbf{q}) = \varepsilon_{ij}^{(0)}(\omega) + j\gamma_{ijk}(\omega) q_k + j\gamma_{ijkm}(\omega) q_k q_m + \cdots \qquad (3.104)$$

Here $\bar{\bar{\mu}} = \mu_0 \bar{\bar{I}}$, and expressions for tensor coefficients $\varepsilon_{ij}^{(0)}(\omega)$, $\gamma_{ijk}(\omega)$ and γ_{ijkm} through covariant tensors b_{ij}, b_{ijk}, b_{ijkl} are evident. The problem how to find EMPs b_{ij}, b_{ijk}, b_{ijkl} refers to the solid-state physics and is beyond the frameworks of our fully classical book.

We only pay attention to the following properties of the triad with components γ_{ijk} and tetrad with components γ_{ijkm}. If $\gamma_{ijk} \equiv 0$ and γ_{ijkm} is symmetric with respect to last two indexes, there is no bianisotropy and no artificial magnetism. This is so in the case of a semiconductor crystal, where the spatial dispersion is fully related with the non-zero phase shift of the eigenwave per unit cell. Although the value $(d/\lambda)^2$ for a natural crystal is extremely small in presence of excitons it cannot be neglected and the term of the second order in (3.104) should be taken into account. Above, we have mentioned that excitons are electron-hole pairs formed in a semiconductor due to the photovoltaic effect and the same term also designates the electromagnetic waves which accompany the movement of an electron-hole pair. More exactly, only longitudinal waves of **E** (similar to the so-called Langmuir waves in a plasma) are related to electron-hole pairs and called exciton waves. Eigenwaves which are neither not similar to the plasma waves also can exist in natural crystals. Among them, there are longitudinal waves without electric and magnetic fields but with nonzero polarization $\mathbf{P} = \mathbf{D} \neq 0$ called mechanical excitons and waves with fully potential **E** and nonzero dipole polarization, called Coulombian excitons.

These manifestations of SD in natural crystals resulting from the retardation of the wave per unit cell combined with the photoelectricity have little to do with our WSD in composite media. Our WSD results mainly from the retardation of the particle response and the non-uniformity of the local electric field across the particle. Our theory obviously requires the homogenization of the microscopic response. As to the effect of retardation of the propagating wave per unit cell, if it is manifested in the electromagnetic response of the medium, it is basically attributed to arrays that cannot be homogenized being effectively discrete. However, sometimes it is not so. In the last chapter of this book we will see that this kind of the retardation effect can be in some special cases also taken into account via the local EMPs and the effective medium is still modeled as a continuous one.

Notice that recently, the approach based on Eq. (3.104) was applied in [182] in order to explain the phenomenon of *negative refraction.* We mentioned this effect in the Introduction (see also in [3]). Negative refraction arises when both permittivity and permeability are negative.

In work [182] it is explained in terms of the second-order SD without involving the permeability. The authors of [182] consider an isotropic non-chiral medium. In this case in (3.104) we have to put $\varepsilon_{ij}^{(0)} = \varepsilon^{(0)} I_{ij}$ and $\gamma_{ijk} \equiv 0$). Next, in [182] an assumption is introduced, expressed by formula (21) of [182], that, in our notations, is the special form of γ_{ijkm}:

$$\gamma_{ijkm}(\omega) = a_1(\omega) I_{ij} I_{km} + \frac{a_2(\omega)}{2}(e_{rik}e_{rjm} + e_{rim}e_{rjk}). \qquad (3.105)$$

This substitution together with $\gamma_{ijk} = 0$ transforms the non-local equation $D_i = \varepsilon_{ij}(\omega, \mathbf{q})E_j$ into a local one:

$$\mathbf{D} = \varepsilon^{(1)}(\omega)\mathbf{E} + a_2(\omega)\nabla \times \nabla \times \mathbf{E}, \qquad (3.106)$$

where it is denoted $\varepsilon^{(1)} = \varepsilon^{(0)} + a_1$. In [182] it is shown—if permittivity $\varepsilon^{(1)}$ is negative and parameter a_2 is sufficiently large the medium exhibits the negative refraction.

On the first glance, this result is something original. However, in fact this sufficiently large positive a_2 can be interpreted as the negative permeability! In fact, let us redefine $\mathbf{D}$ and $\mathbf{H}$ substituting $\mathbf{T} = -a_2 \nabla \times \mathbf{E}$ into relation (3.80). Then we obtain

$$\mathbf{D} = \varepsilon^{(1)}(\omega)\mathbf{E}, \quad \mathbf{B} = \mu(\omega)\mathbf{H}, \qquad (3.107)$$

where it is denoted

$$\mu(\omega) = \frac{\mu_0}{1 - \omega^2 \mu_0 a_2}. \qquad (3.108)$$

Formulas (3.107) and (3.108) transform MEs of the medium with second-order SD to MEs of a usual isotropic magneto-dielectric. This redefinition does not violate the covariance of material parameters since all the parameters in the initial equation (3.104) were introduced in the covariant form. Notice that formula (3.108) for isotropic non-chiral media with SD of the second order was derived in book [79].

Therefore, formula (3.105) from [182] with the conditions of the negative refraction $\varepsilon^{(1)} < 0$ and $\omega^2 \mu_0 a_2 > 1$ are equivalent to the assumption of local isotropic permittivity and permeability in the medium without chirality and b-parameter (in which the term grad$\nabla \cdot \mathbf{E}$ in Eq. (3.94) vanishes). In other words, paper [182] gives the same conditions for the negative refraction as were obtained in [3]. The second-order term of the non-local permittivity results in the non-trivial permeability.

Notice that an anisotropic analogue of formula (3.108) was derived in [183]:

$$\mu_{ii} = \mu_0 \left(1 - \frac{\omega^2 \mu_0}{2} \frac{\partial^2 \varepsilon_{jj}(\omega, \mathbf{q})}{\partial q_k^2}\bigg|_{q_i=q_j=q_k=0}\right)^{-1}. \quad (3.109)$$

In (3.109) the permittivity with components ε_{jj} is defined by Eq. (3.104). Above, we have presented the material parameters in an arbitrary Cartesian system where they could comprise off-diagonal terms. Relation (3.109) was strictly derived for a lossless (practically, low-loss) lattice. In this case the tensor $\bar{\bar{\varepsilon}}$ is diagonal if expressed in the lattice coordinate system. The model of the homogenization of lattices discussed in [19] also neglects all higher multipoles. This work allows one to understand better the bounds between weak and strong spatial dispersion and also explains the usefulness of non-local material parameters for special cases which were already discussed above.

3.6 Some Preliminary Conclusions

Our theory reveals the following features of the reciprocal composites (also inherent to media of complex-shape molecules):

- The bianisotropy of a linear reciprocal medium also called the medium magneto-electric coupling and comprising the chirality as a special case is the effect of SD of the first order. The artificial magnetism described by the medium permeability is the effect of SD of the second order.
- If the eigenwave of the medium induces in its particles the high-order multipoles both magneto-electric coupling parameter and permeability contain contributions from multipolar susceptibilities additionally to the electric and magnetic dipole susceptibilities. The electric quadrupole susceptibility contributes into the MEC parameter. Several multipole susceptibilities contribute into the permittivity and permeability.

- In media with SD of the second order (except the special case when the high-order multipoles are absent) the first material equation contains the second-order derivatives of **E**. Besides permittivity, permeability, and MEC parameter, there is the fourth tensor (b-parameter) corresponding to the second-order spatial derivatives of the mean field. This parameter is important for an effective medium composed by resonant dimers and for dense arrays of substantially long and thin metal needles.

These conclusions should force us to critically revise numerous scientific publications devoted to MMs, especially those in which optical nanostructured MMs with resonant magnetic response are characterized in terms of only effective permittivity and permeability.

How to understand qualitatively which of the multipole susceptibilities are important and which can be neglected for an explicit array of scatterers? This understanding can be achieved in the qualitative analysis of the individual constitutive particle to the non-uniform local field. In the next section we give an example of this analysis for a composite of parallel Π-shaped particles performed of a thin metal wire.

3.7 How the Non-Locality of the Particle Response Results in WSD

3.7.1 A Qualitative Microscopic Analysis of an Explicit Example

Consider an optically small particle of complex shape whose material is strongly contrasting with the ambient, e.g., a bent/folded metal wire in a dielectric host. Namely, let it be a Π-shaped particle depicted in Fig. 3.4. Locate the coordinate frame so that the particle plane is $(x - y)$. Then only x and y components of the local field $\mathbf{E}^{loc}$ are essential for its excitation. Polarization current $\mathbf{J}^{micro}$ is induced in the particle by the non-uniform local field $\mathbf{E}^{loc}$. The most general form of the relation between $\mathbf{J}^{micro}$ and $\mathbf{E}^{loc}$ that takes into account both retardation of the coupling between different parts of the particle and the non-uniformity of $\mathbf{E}^{loc}$ is as follows (in Cartesian coordinates):

$$J_i^{micro}(\mathbf{R}) = \int_\Omega K_{ij}^{micro}(|\mathbf{R} - \mathbf{R}'|) E_j^{loc}(\mathbf{R}') dV'. \qquad (3.110)$$

Here Ω is the particle volume.

Dipole moment $\mathbf{p}$ of the particle has the following Cartesian components:

$$p_i = \frac{1}{j\omega} \int_\Omega \int_\Omega K_{ij}^{micro}(|\mathbf{R} - \mathbf{R}'|) E_j^{loc}(\mathbf{R}') dV' dV. \qquad (3.111)$$

In accordance with the reciprocity principle, the dyad $\overline{\overline{K}}$ is symmetric: $K_{ij}^{micro} = K_{ji}^{micro}$ and depends only on the distance $|\mathbf{R}-\mathbf{R}'|$ between the integration and observation points (see, e.g., [138]).[f]

Expand the local field at the arbitrary integration point $\mathbf{R}'$ into the Taylor series with respect to the coordinate origin—a certain reasonably chosen point of the medium unit cell to which we refer the dipole $\mathbf{p}$ and all other multipoles of the particle. The assumption of the particle smallness allows us to keep only two terms:

$$E_j^{loc}(\mathbf{R}') = E_j^{loc}\Big|_0 + (\nabla_k E_j^{loc})\Big|_0 R_k' + \frac{1}{2}(\nabla_l \nabla_k E_j^{loc})\Big|_0 R_k' R_l' \cdots \qquad (3.112)$$

Substituting expansion (3.112) into (3.111) and taking into account that the dipole moment $\mathbf{p}$ and medium polarization at the coordinate origin are related via the unit cell size d as $p_i = P_i d^3$, we obtain:

$$P_i(\mathbf{r}) = \alpha_{ij} E_j^{loc}\Big|_\mathbf{r} + \alpha'_{ijk}(\nabla_k E_j^{loc})\Big|_\mathbf{r} + \frac{1}{2}\alpha''_{ijkl}(\nabla_l \nabla_k E_j^{loc})\Big|_\mathbf{r} \cdots, \qquad (3.113)$$

where we have for the dyad α_{ij} (do not mix with the dyad a_{ij} relating the polarization to the mean field):

$$\alpha_{ij} = \frac{1}{j\omega d^3} \int_\Omega \int_\Omega K_{ij}^{micro}(|\mathbf{R} - \mathbf{R}'|) dV' dV, \qquad (3.114)$$

---

[f] In fact, K_{ij}^{micro} can be expressed through the dyadic Green function of the scattering particle. However, this relation is out of the scope of our current study.

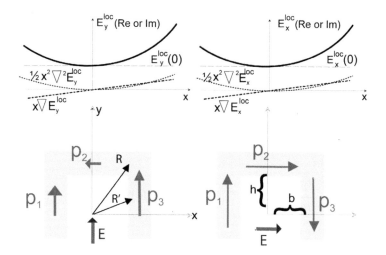

Figure 3.4 A Π-shaped metal particle in the nonuniform local field. On the top panel both essential components of E^{loc} are decomposed onto three parts corresponding to the second-order theory—uniform, linear and quadratic parts. Left panel—E_y^{loc} induces a cross-polarized p-dipole, a m-dipole and a quadrupole due to the linear part of its x dependence. Right panel—uniform E_x^{loc} induces a m-dipole and a quadrupole.

and for the triad α_{ijk} (do not mix with the tried α'_{ijk} referring to the derivatives of the mean field):

$$\alpha'_{ijk} = \frac{1}{j\omega d^3} \int_\Omega \int_\Omega K_{ij}^{\text{micro}}(|\mathbf{R} - \mathbf{R}'|) R'_k \, dV' dV \qquad (3.115)$$

The expression for the tetrad α''_{ijkl} can be also easily written but we do not need it.

The first term in (3.113) is the response of the unit cell to the uniform part of the local field ($E_x^{loc}(0)$ and $E_y^{loc}(0)$ in Fig. 3.4). The second term is the response to the linearly varying part of the local field ($x \nabla E_x^{loc}(0)$ and $x \nabla E_y^{loc}(0)$ in Fig. 3.4). The third term is the response to the quadratic part of the local field ($0.5 x^2 \nabla^2 E_x^{loc}(0)$ and $0.5 x^2 \nabla^2 E_y^{loc}(0)$ in Fig. 3.4). In this interpretation the parameter of smallness is not the optical size of the unit cell qd, but the optical size of the particle qa. However, in practice they are two values of the same order.[g]

In the expression (3.113) one can explicitly share out the contribution of the local magnetic field. For it the tensor α'_{ijk} is decomposed onto a symmetric and antisymmetric parts:

$$\delta_{ijk} = \frac{\alpha'_{ijk} + \alpha'_{ikj}}{2}, \quad u_{ijk} = \frac{\alpha'_{ijk} - \alpha'_{ikj}}{2}. \quad (3.116)$$

Like we did above, the non-symmetric part $\overline{\overline{\overline{u}}}$ represents a product of a certain dyad we can denote as $(\overline{\overline{\gamma}}/j\omega)$ by a Levi–Civita tensor: $\overline{\overline{\overline{u}}} = (j\omega)^{-1}\overline{\overline{\gamma}} \cdot \overline{\overline{\overline{e}}}$. The product of the Levi–Civita tensor triad by ∇ gives the curl operator $\nabla\times$. We understand that microscopic Maxwell's equations hold for partial fields created by whatever sources located in this medium. Therefore, they keep valid for local fields and we may apply Eq. (3.31) for local fields as well—it does not contain the medium response. The substitution of (3.31) transforms (3.113) into

$$P_i = \alpha_{ij} E_j^{\text{loc}} + \delta_{ijk}(\nabla_k E_j^{\text{loc}}) - \gamma_{ij} B_j^{\text{loc}} + \frac{1}{2}\alpha''_{ijkl}(\nabla_l \nabla_k E_j^{\text{loc}}) + \cdots \quad (3.117)$$

Once more, we see that the response of the dipole moment to the anti-symmetric part of the local field (of type $x\partial E_y/\partial x$ or $y\partial E_x/\partial y$, etc.) can be interpreted as the response to the local magnetic field. This response implies the bianisotropy.

In order to better understand the qualitative importance of the anti-symmetric term in $\mathbf{E}^{\text{loc}}$ let us look at Fig. 3.4 and split the particle onto 3 elements—stem 1, crossbar 2 and stem 3. Consider the response of this particle to the non-uniform local field $E_y^{\text{loc}}(x)$ modeled in accordance with the second-order theory as a quadratic polynomial. The uniform part $E_y^{\text{loc}}(0)$ and the quadratic part $0.5x^2 \nabla^2 E_y^{\text{loc}}(0)$ both create equivalent dipole moments in stems 1 and 3 that we may denote as $\mathbf{p}_0$. The anti-symmetric part $x \nabla E_y^{\text{loc}}(0)$ creates in stems 1 and 3 the opposite dipole moments,

---

[g] In our theory d is smaller than λ by an order of magnitude. If also a is smaller than d by an order of magnitude, we have $(a/\lambda) \sim 0.01$. Neither a ceramic particle with permittivity 100 nor a metal particle of whatever shape can have a noticeable magnetic response at such a wavelength. In other words, WSD is not observed in geometrically sparse composites. The particle size a is noticeably smaller than the structural period d, as we have discussed above, but not by an order of magnitude. Practically, $d = (1.5\cdots 3)a$.

that we may denote as $-\Delta\mathbf{p}$ and $+\Delta\mathbf{p}$, respectively. As a result, we have in stem 1 the dipole $\mathbf{p}_1 = \mathbf{p}_0 - \Delta\mathbf{p}$ and in stem 3 the dipole $\mathbf{p}_3 = \mathbf{p}_0 + \Delta\mathbf{p}$. Since $\mathbf{p}_3 \neq \mathbf{p}_1$ the current in the crossbar between the stems arises and results in the cross-polarization $\mathbf{p}_2$. The response of the particle is a combination of two y-directed dipoles $\mathbf{p}_1$ and $\mathbf{p}_3$ and one x-directed dipole $\mathbf{p}_2$. The total dipole moment $\mathbf{p} = \mathbf{p}_1 + \mathbf{p}_2 + \mathbf{p}_3$ results from the uniform and quadratic parts of $E_y^{loc}(x)$ whose contributions are not distinguishable from one another. If the local field is along y there is no MEC—$E_y^{loc}(0)$ does not induced the m-dipole. Is there the artificial magnetism for this polarization?

Yes, it is. The magnetic moment referred to the particle center $(0, 0)$ can be expressed through the dipole moments of the particle segments. In accordance with the general definition of the magnetic moment (3.20), we have: $m = m_z = j\omega\mu_0[(p_1 - p_3)b + p_2 h]$. It is fully induced by the anti-symmetric part of the local electric field $x\partial E_y/\partial x$, i.e., is proportional to the cross-derivative of the local field. In other words, the magnetic moment results from the z-directed local magnetic field B_z^{loc}. The same refers to the quadrupole moment which has two components q_{xy} and q_{yx}. This is how we can estimate the individual polarizabilities entering (3.117) in the case when $\mathbf{E}^{loc}$ is polarized along y. The case when $E^{loc} = E_y^{loc}$ and the wave propagates along z is trivial—the variation of the field along z produces nothing and the particle is a simple electric dipole. The propagation along x is the most important case—then the artificial magnetism and bianisotropy arise together.

Now, let us consider the response to the non-uniform x-polarized field $E_x^{loc}(x)$. Then the uniform and quadratic terms of this field produce the dipole $\mathbf{p}_2$ in the crossbar. The corresponding x-directed polarization current is continued in the stems by y-polarized currents, that correspond to the dipoles $\mathbf{p}_1$ and $\mathbf{p}_3 = -\mathbf{p}_1$ in the stems 1 and 3, respectively. Opposite signs of the p-dipoles of stems 1 and 3 mean that the uniform electric field produces the z-directed magnetic moment $m = j\omega\mu_0(2p_1 b + p_2 h)$ and a two-component quadrupole moment. This is the MEC—the bianisotropy of the particle response. As to the action of the anti-symmetric term $x\nabla E_x^{loc}(0)$, it is evidently absent.

So, besides the electric polarization (that is common with the action of $E_y^{loc}(x)$) the uniform E_x^{loc} produces the first-order effect of SD—the bianisotropy. There is no artificial magnetism related with E_x^{loc}. It means that the magnetic response is absent if the wave propagates along y. And, of course, three dipoles $\mathbf{p}_1$, $\mathbf{p}_2$, and $\mathbf{p}_3$ form also a magnetic quadrupole response that is responsible for the nonzero b-parameter. However, this parameter keeps rather small at the frequencies where the artificial magnetism and bianisotropy are resonant. At these frequencies the total length of the wire $l_{tot} = 4h + 2b$ is close to one half of the wavelength $\lambda/2$. For a Π-shaped particle of nearly perfectly conducting wire (operating at microwaves) these two resonance bands may be separated over the frequency axis. In the optical case when the particle is performed of a plasmonic metal nanowire, the resonances of the magnetic and the BA responses overlap and the MEC parameter prevails. In the resonance band the characteristic size of the particle $\max(2h, 2b)$ is nearly six times smaller than λ. This means that the composite of these particles like a composite of small resonant dimers has the prerequisites to be effectively continuous. However, it will be not a simple BA medium of p-m dipoles like that of OPs because in its MEC parameter the electric quadrupole susceptibility is present.

In Fig. 3.4 the case when the electric field varies along y is not shown. However, it is rather clear that the uniform and quadratic parts of E_x^{loc} again induce the p-dipole in the crossbar and a quadrupole together with the m-dipole in the stems, and E_y^{loc} induces nothing but the p-dipole. The crucial difference of this case from the case when the field varies along x is that the MEC parameter is zero, and the medium behaves as a simple dielectric. The m-dipole and the quadrupole polarizabilities in the individual MEC parameter γ cancel out and the BA response is absent. It also results from the generalized Maxwell Garnett model, developed below, but it is clear in advance from this microscopic qualitative analysis. The opposite p-dipole moments in the stems cannot radiate along y and do not contribute into the wave inducing such a polarization in the particle. The z-directed m-dipole related with these two dipoles and the xy-component of the quadrupole moment related with them cancel one another in the fields calculated outside the particle.

The reader can see from this consideration that a Π-shaped particle is not a split-ring resonator as it is assumed in the abundant literature. Media formed by such arrays are not simply magneto-dielectrics, as one could conclude from papers [175–179]. For one sheer of propagation directions it is a simple dielectric. For another sheer of propagation directions it is a BA medium. In such a BA medium operating at optical frequencies the MEC parameter prevails compared to the artificial magnetism. The magnetic moment of a unit cell of the corresponding lattice attributed in papers [175–179] to the resonant effective permeability refers mainly to the resonant MEC parameter.

3.7.2 Spatial Dispersion Expressed by the b-Parameter

The understanding of the WSD as a phenomenon is based on the possibility to expand the non-uniform local electric field into the rapidly converging Taylor series around the multipole center of the constitutive particle. Only three terms in this expansion—uniform, linear and quadratic—are retained. This restriction is adequate because the particle is optically small. The account of the linear and quadratic terms is relevant because the optically small resonant particle can have the complex shape when made of a metal wire or can experience the so-called Mie resonance when made of a high-permittivity dielectric.

The m-dipole and the electric quadrupole may both respond to the antisymmetric derivatives (cross-derivatives) of the local field, i.e., to the spatial derivatives of type $\partial E_x/\partial y$. The response to the m-dipole and electric quadrupole to these field derivatives is their response to the local magnetic field called artificial magnetism.

Only the off-diagonal components of the electric quadrupole are excited by this part of the local field, and this excitation brings no new physics compared to the magnetization. The quadrupole and magnetic dipole susceptibilities enter into MEs in a covariant combination and are therefore applicable to any observation point.

Besides the bianisotropy and artificial magnetism, there is another manifestation of WSD in the second order of smallness—that resulting in the so-called b-parameter. This parameter reflects the response of the p-dipole to the quadratic part of the mean field and the rise of the electric octupole. These two effects can be united into one effect of the b-parameter since they hold for substantial particles due to the noticeable retardation of the local field in them. How important is the b-parameter effect?

In effectively continuous media b-parameter cannot be resonant, it becomes resonant at frequencies where the particle overall size is resonant. In this case the medium whose unit cell comprises one particle is not optically dense and obviously acquires the strong SD. However, we may imagine the situation when our theory is still applicable but the medium cannot be split to unit cells comprising exactly one particle.

Imagine that our composite is formed by thin and rather long (but shorter than $\lambda/2$) metal needles. The needles can be arranged in parallel and even can form an isotropic mixture without touching one another. In both cases the distances between the centers of the adjacent needs are much smaller than their lengths and the composite is optically dense. If the density of needles is high, the capacitive coupling between the adjacent needles is strong. Then the medium is not penetrable for the electromagnetic wave, as we have discussed above. If the volume fraction of needs is small enough and the medium is penetrable for the electromagnetic wave, it is the case when our theory is applicable and the b-parameter is important. If the needles are all oriented in parallel they form an effectively homogeneous composite with a very strong anisotropy manifested in both permittivity and b-parameter. An isotropic sufficiently sparse mixture of needles is characterized by the isotropic b-parameter, when MEs (3.94) and (3.95) take form

$$\mathbf{D} = \varepsilon \mathbf{E} + b \nabla \nabla \cdot \mathbf{E}, \quad \mathbf{B} = \mu_0 \mathbf{H}.$$

Besides an effectively continuous medium of needs, the b-parameter can be important for effectively continuous medium of dimers or oligomers—or a complex-shaped particle with non-closed loops such as a S-shaped one, C-shaped one and Π-shaped one. Such media can be effectively continuous in the band of the fundamental resonance of the constitutive particles. It is clear that in the resonance band the b-parameter can be also resonant and therefore is not negligible.

3.7.3 A Relation between the Dipole and the Quadrupole Susceptibilities

Now, as it was promised above, let us derive the relation (3.49). This relation as well as its magnetic analogue (3.78) has a key importance for the theory of WSD. Without (3.49) and (3.78) the covariant MEs would be impossible for multipolar media. We will, first, derive these relations on the microscopic level for the polarizabilities of a particle. Then, using the CMLL formula we will generalize them to the medium susceptibilities.

Although the CMLL formula is valid for the particle centers, any point of a unit cell can be considered as the corresponding particle center. Therefore, we we may write it down for an arbitrary point **R** located inside the array of particles. In the index form the anisotropic the CMLL formula corresponding to Eq. (2.14) is as follows:

$$E_i^{\text{loc}}(\mathbf{R}) = E_i(\mathbf{R}) + N_{ij} P_j(\mathbf{R}), \qquad (3.118)$$

where the dyad $\overline{\overline{N}}$ does not depend on the observation point location and on the parameters of the wave polarizing the particles.[h] Substituting (3.118) into (3.117), we obtain for the mean p-dipole polarization

$$P_i = \alpha_{ij}(E_j + N_{ij} P_j) + \frac{1}{2}\delta_{ijk}\nabla_k(E_j + N_{ij} P_j) + \gamma_{ij} B_j + \cdots. \quad (3.119)$$

Here, in accordance with (3.115) and (3.116), we have

$$\delta_{ijk} = \frac{1}{j\omega d^3} \int_\Omega \int_\Omega (K_{ij}^{\text{micro}} R'_k + K_{ik}^{\text{micro}} R'_j) dV' dV. \qquad (3.120)$$

Relationship (3.120) is symmetric with respect to commutations $\mathbf{R} \to \mathbf{R}'$ and $\mathbf{R} \to \mathbf{R}'$.

Now, let us inspect the triad $\overline{\overline{d}}$ (the medium p-dipole susceptibility to the symmetric spatial derivatives of the mean field **E**) and

---

[h]In the special case of a cubic lattice or a random mixture it degenerates into $N = 1/3\varepsilon_m\varepsilon_0$.

the triad $\overline{\overline{\overline{Q}}}$ (the medium quadrupole susceptibility to **E**). Following to the relationship $Q_{ik} = q_{ik}/\Omega$ between the particle quadrupole moment q_{ik} and the mean quadrupole density Q_{ik} and substituting formula (3.110) into the definition (3.18) of the electric quadrupole, we obtain:

$$Q_{ik} = \frac{1}{j\omega d^3} \int_\Omega \int_\Omega \left(K_{ij}^{\text{micro}}(|\mathbf{R}-\mathbf{R}'|)R_k + K_{kj}^{\text{micro}}(|\mathbf{R}-\mathbf{R}'|)R_i\right) \cdot$$

$$E_j^{\text{loc}}(\mathbf{R}')\,dV'dV. \qquad (3.121)$$

The coefficient Q'_{ijk} (we want to express it through $\overline{\overline{\overline{d}}}$) is the value of the first order of smallness. Therefore, substituting the expansion (3.112) for the local field into (3.121), we may restrict the expansion by the first-order terms. Substituting formula (3.118) into (3.121), we obtain for the quadrupole moment a first-order formula in which the dipole susceptibility to the symmetric derivatives to the mean field enter:

$$Q_{ij} = \beta_{ijk}(E_k + N_{ij}P_k), \qquad (3.122)$$

where we have denoted

$$\beta_{ijk} \equiv \frac{1}{j\omega d^3} \int_\Omega \int_\Omega \left(K_{ij}^{\text{micro}} R_k + K_{jk}^{\text{micro}} R_i\right) dV'dV = \delta_{jki}. \qquad (3.123)$$

The equivalence in (3.123) evidently follows from (3.120). Here for simplicity of writing we have replaced the tensor f_{ij} by a scalar f which will not enter the final result.

For further simplicity, let us assume that in (3.119) the magneto-electric polarizability $\overline{\overline{\gamma}}$ is equal to zero. This assumption does not change the result, since the relationship we are looking for, has nothing to do with the bianisotropy. This assumption only makes all formulas shorter. If in the expansion (3.119) we keep only terms of the zero order, we have

$$P_i = (I_{im} - N_{ij}\alpha_{jm})^{-1}\alpha_{ml}E_l, \qquad (3.124)$$

where $\overline{\overline{I}}$ is a unit dyad. Since the quadrupole moment itself is the first-order value, when we substitute **P** into (3.121) we have to use namely Eq. (3.124). Then we have two relations in which only zero-order and first-order terms are present:

$$Q_{ij} = \beta_{ijk}E_k + N_{is}\beta_{jsk}(I_{km} - f\alpha_{km})^{-1}\alpha_{lm}E_l, \qquad (3.125)$$

and

$$P_i = (I_{ik} - N_{is}\alpha_{sk})^{-1}\alpha_{kj} E_j + \frac{\delta_{ijk}}{2}\nabla_k E_j$$
$$+ \frac{\delta_{ijk}}{2} N_{sp}(I_{sp} - N_{sq}\alpha_{qm})^{-1}\alpha_{ml}\nabla_k E_l. \quad (3.126)$$

After a simple algebra these relations read as follows:

$$Q_{ij} = Q'_{ijm} E_m$$
$$= \beta_{ijk}\left[I_{km} + N_{kp}\alpha_{ps}(I_{sm} - N_{sq}\alpha_{mq})^{-1}\right] E_m. \quad (3.127)$$

and

$$P_i = (I_{ik} - N_{iq}\alpha_{qk})^{-1}\alpha_{kj} E_j + \frac{\delta_{ijk}}{2}\left[I_{jl} + N_{jq}\alpha_{ql}(I_{jm} - f\alpha_{jm})^{-1}\right]\nabla_k E_l. \quad (3.128)$$

In accordance with (3.122) $\beta_{ijk} = \delta_{jki} = \delta_{jik}$, and in accordance with the reciprocity $\alpha_{ml} = \alpha_{lm}$. Therefore, Eq. (3.127) can be rewritten as

$$Q'_{ijm} = \delta_{ijk}\left[I_{km} + N_{kq}\alpha_{qm}(I_{qm} - N_{ql}\alpha_{ml})^{-1}\right]. \quad (3.129)$$

Now let us recall Eq. (3.43) relating the mean p-dipole polarization with the mean field and its spatial derivatives. Substituting (3.48) into (3.43), and taking into account that in the case $\bar{\bar{\gamma}} = 0$ there is no bianisotropy and the tensor $\bar{\bar{g}}$ also vanishes,[i] we obtain

$$P_i = a_{im} E_m + \frac{d_{ijk}}{2}\nabla_k E_j. \quad (3.130)$$

Equating the right-hand side of (3.130) to that of (3.128) we have in the first order of smallness

$$d_{ikl} = d_{ilk} = \delta_{ikj}\left[I_{jl} + N_{jq}\alpha_{ml}(I_{qm} - N_{qs}\alpha_{sm})^{-1}\right]. \quad (3.131)$$

Of course, equating (3.130) to (3.128) in the zero order we will also relate the dipole medium susceptibility with the microscopic polarizability, however, in not the purpose of the present derivation. Moreover, this formula does not differ from the standard Lorentz-Lorenz one because we have ignored the bianisotropy.

What we need is to compare relations (3.131) and (3.129) and see that their right-hand sides coincide because j and m in (3.131) are only repeating indexes of summation. Replacing in the notations

---

[i] It can be proved strictly, but we avoid these calculations for brevity.

of (3.129) j by k and m by l we finally obtain the coincidence of the right-hand sides of (3.129) and (3.131):

$$Q'_{ikl} = Q'_{kil} = d_{lik} = d_{lki}.$$

Thus, relations (3.49) are proved. Note that these formulas are also correct for multilevel quantum emitters in which both dipole and quadrupole optical transitions exist [165]. Equations (3.49), as well as their analogues Eqs. (3.78), are global relations and really deserve the attention we have paid.

Also, our derivations were an instructive illustration of the following statement. If there is an equation unambiguously relating the mean field with the local field in the zero order of smallness, all EMPs can be expressed through the individual particle response. In general, it is the function $\overline{\overline{K}}^{micro}(\mathbf{R})$ relating the induced polarization current with the distribution of the non-uniform local electric field.

In order to share the particle polarizabilities from the numerical simulations it is instructive to combine the incident plane waves so that to find the multipolar response to the local electric field and to its cross-derivative at the particle center. In the first case we should center the particle at the magnetic node of the standing wave, and in the second case—in the electric node. As a rule, the particle possess a certain symmetry and the response to the magnetic field polarized along different coordinate axes may be related with the response to the cross derivatives of the electric field $\mathbf{E}$. Ones we have found the particle responses to the uniform part of $\mathbf{E}$ and to its cross derivatives, the response to the symmetric derivatives of $\mathbf{E}$ can be extracted from the simulation for the case when two incident waves propagates with a sharp or obtuse angle to one another. In fact, exciting a particle by different pairs of coherent plane waves one may find all multipolar polarizabilities involved into the theory of WSD.

3.7.4 On the Hierarchy of Multipoles for Densely Packed Media

The hierarchy of multipoles in the theory of WSD, where qd is parameter of smallness, implies that in the constitutive particles the following effects of the first order arise:

- the pair m-dipole plus electric quadrupole responding to the local electric field at the particle center;
- the p-dipole acquires the part responding to the first derivative of the local electric field at the particle center.

Both these effects are manifested as the bianisotropy. Antisymmetric first derivative of the local electric field can be replaced by the local magnetic field.

In the second order, the following effects arise:

- the pair magnetic quadrupole plus electric octupole responding to the local electric field at the center;
- the p-dipole acquires the part responding to the second derivative of $\mathbf{E}^{loc}$;
- the m-dipole and electric quadrupole acquire the part responding to the first derivative of $\mathbf{E}^{loc}$.

Our homogenization model based on the multipole hierarchy implies that the second-order response parameters are smaller than the parameters of the first order and the third-order and higher-orders parameters are negligible. It was already noticed that this hierarchy may be destroyed by the multipole resonances, e.g., by that of the m-dipole that may arise in the frequency band where the medium is still effectively continuous. For the band of the m-dipole resonance the Maxwell Garnett homogenization model generalized for media with WSD in the end of this section will be not valid.

Here, it is worth to notice that our hierarchy also destroys if the particle response to the static field ($qd = 0$) is not a p-dipole. In fact, high-order electric multipoles may arise in a particle due to the non-uniformity of the local *static* field. This non-uniformity ignored in our model has nothing to do with the retardation of the medium eigenwave over the particle that is a prerequisite of WSD.

When can the local field be essentially non-uniform in the static limit? We do not consider the embedded sources, but it also can occur if the particles are densely packed. Then the field produced by the adjacent particles inside the reference one is strongly not uniform even in statics. If the internal field is essentially non-uniform, the static electric multipoles arise in the response of this particle. For such a structure the multipole hierarchy does not hold, and the point-to-point model of the electromagnetic interaction between particles is not correct.

We have already claimed that for the dense arrays of complex-shaped metal particles our theory is not applicable and we are not interested in this case—such a medium is weakly transparent. The last comment is even more convincing for simple-shaped metal particles. If the metal spheres or cubes are almost touching the effective medium is nearly opaque at low frequencies where it can be homogenized.

However, for all-dielectric composites we allowed the case of the dense package. How dense can it be so that our model keeps applicability? Lord Rayleigh in [184] formulated the applicability of the dipole approximation as follows: "The distance between the elements of an array should be sufficient so that the polarization response of each element to the field applied to it from all other polarized elements and external sources could be approximated as uniform within the element."

For dielectric spheres forming a simple cubic lattice the static multipoles can be neglected even if the spheres are mutually touching. For the dense arrangement such as a face- or body-centered cubic lattice one needs a certain gap between the adjacent spheres. For metal spheres forming a simple cubic lattice the needed gap should be at least 15–20% of the lattice period [112]. This is always respected in arrays of plasmonic nanospheres, otherwise the deviations due to fabrication tolerances may result in the galvanic contact of the spheres. For complex-shape particles the negligence of the static multipoles is more demanding to the separation of particles. To neglect the static multipoles in arrays of canonical CPs or OPs one needs the separation between the closest edges of the neighboring particles 30–50% of the period [150, 151, 185]. The last restriction has been already introduced above as a bound of applicability of the whole theory. So, the requirement of the absence of static multipoles as a prerequisite of the multipole hierarchy does not bring new restrictions to the geometries of the homogenized arrays. These restrictions have been already formulated above.

3.7.5 Generalized Maxwell Garnett Model for Media with WSD

The simplest version of the generalized Maxwell Garnett model—that applicable to the case when the particles do not possess the quadrupole response—was already presented in Chapter 1. In the present chapter we consider the general case of WSD—a multipolar response—but for simplicity assume that the array has the isotropic geometry (cubic lattice or its slightly irregular variant). Then the CMLL formula for the electric field (2.4) that we present here in the form

$$\mathbf{E}^{loc} = \mathbf{E} + \frac{1}{3\varepsilon_0 \varepsilon_m} \mathbf{P}, \tag{3.132}$$

taking into account the matrix permittivity, and its magnetic analogue (2.17) that we rewrite here as

$$\mathbf{B}^{loc} = \mathbf{B} + \frac{\mu_0}{3} \mathbf{M}, \tag{3.133}$$

taking into account the definition of the magnetic moment, adopted in SI, do not contain tensors.

In this subsection we will derive the Maxwell Garnett formulas based on the definitions of EMPs we have obtained deriving our MEs (3.86), (3.87), and on Eqs. (3.132), (3.133).

For simplicity we reproduce only the derivations corresponding to the first-order theory. The mixing rule for the second-order EMP will be given without derivations. Here we cover only artificial magnetism. For the b-parameter the mixing rule was not derived: the arrays for which this EMP is significant are not interesting for the author of the present book.

In the first-order approximation formula (3.117) takes form:

$$p_i \equiv \frac{P_i}{n} = \alpha_{ij} E_j^{loc} + \frac{1}{2}\delta_{ijk}\nabla_j E_k^{loc} - j\omega\gamma_{ij} B_j^{loc}. \tag{3.134}$$

Here the microscopic polarizablities α_{ij}, δ_{ijk} and γ_{ij} are assumed to be known. Also, we have in mind that our theory refers to rather dilute composites. Therefore, the terms proportional to high powers of the particles concentration, such as n^2, n^3, etc., will be relatively small compared to the terms proportional to n.

We have already discussed above how to modify the model if there are deviations in the particles orientations. Writing the relation $p_i \equiv P_i/n$ we assume by default that all particles are oriented in parallel.

Using the dyadic form we may write the result of the substitution of (3.132) and (3.133) into (3.134) in a form:

$$\left(\overline{\overline{I}} - \frac{n\overline{\overline{\alpha}}}{3\epsilon_0\varepsilon_m}\right) \cdot \frac{\mathbf{P}}{n} = \overline{\overline{\alpha}} \cdot \mathbf{E} + \frac{1}{2}(\overline{\overline{\delta}} \cdot \nabla) \cdot (\mathbf{E} + \frac{\mathbf{P}}{3\epsilon_0\varepsilon_m}) - j\omega\overline{\overline{\gamma}} \cdot \mathbf{B} - \frac{j\omega}{3}\overline{\overline{\gamma}} \cdot \mathbf{M}. \tag{3.135}$$

So, the dipole polarization of the medium is expressed now through mean electric and magnetic fields and magnetic polarization.

Consider the expression for the magnetic moment of a particle

$$m_i = m'_{ij} E_j^{\text{loc}} \tag{3.136}$$

following from (3.45) in the first-order theory. Due to the reciprocity the dyadic function $\overline{\overline{K}}^{\text{micro}}$ describing the particle response to the non-uniform local field is a symmetric dyad: $K_{ij}^{\text{micro}} = K_{ji}^{\text{micro}}$. From (3.110) and (3.115), we have $j\omega\gamma_{ij} = m'_{ji}$, or, equivalently,

$$j\omega\overline{\overline{\gamma}} = (\overline{\overline{m}}')^T. \tag{3.137}$$

We have already mentioned this formula called in the theory of scattering the molecular reciprocity theorem. It was independently derived in [138] and in [169] from the Lorentz lemma. Here we have used the reciprocity of the kernel $\overline{\overline{K}}$, that is an equivalent approach.

Formula (3.137) is the microscopic analogue of the corresponding macroscopic reciprocity (3.72). Equation (3.72) was imposed in [80] as a necessary condition of reciprocity of any BA medium (the sufficient condition of reciprocity implies that (3.72) is complemented by the symmetry of the permittivity and permeability tensors). However, to impose such conditions is a phenomenological approach. We cannot *require* the reciprocity of our effective medium. We build its model and have to prove that it is built correctly. In particular, we have to prove that our model keeps the effective medium reciprocal if its constituents are reciprocal.

The magnetic moment is the first-order value. Consequently, in the right-hand side of (3.136) only terms of the zero order should be taken into account. Substituting (3.132) and (3.136) into (3.134) we have to use only the zero-order terms (as we did above deriving the relation (3.49) between the dipole differential and quadrupole integral susceptibilities of the medium). Therefore, (3.135) simplifies to

$$\frac{\mathbf{P}}{n} = \left(\overline{\overline{I}} - \frac{n\overline{\overline{\alpha}}}{3\epsilon_0\varepsilon_m}\right)^{-1} \cdot \overline{\overline{\alpha}} \cdot \mathbf{E}. \tag{3.138}$$

Both dyadic factors in the right-hand side of (3.138) are symmetric.

For the magnetic moment we similarly obtain from (3.136):

$$\frac{\mathbf{M}}{n} = \overline{\overline{m}}' \cdot \left[\overline{\overline{I}} + \frac{n\overline{\overline{\alpha}}}{3\epsilon_0\varepsilon_m} \cdot \left(\overline{\overline{I}} - \frac{n\overline{\overline{\alpha}}}{3\epsilon_0\varepsilon_m}\right)^{-1}\right] \cdot \mathbf{E}. \tag{3.139}$$

After the multiplication in the right-hand side of (3.139) we earn the terms of the order n and n^2. The term proportional to n is that of the order $(a/d)^3$, and the term with n^2 is that of the order $(a/d)^6$. Since for complex-shape particles (otherwise there is no resonant quadrupole response) the theory of WSD requires $d > 1.5a$ (see above), the term with n^2 is negligible. Skipping it we obtain from (3.139):

$$\mathbf{M} = n\overline{\overline{m}}' \cdot \left(\overline{\overline{I}} - \frac{n\overline{\overline{\alpha}}}{3\epsilon_0\varepsilon_m}\right)^{-1} \cdot \mathbf{E}. \tag{3.140}$$

The expression (3.140) can be substituted into (3.135), taking into account that $j\omega\overline{\overline{\gamma}} = (\overline{\overline{m}}')^T$. These substitutions give the following formula:

$$\frac{\mathbf{P}}{n} \cdot (\overline{\overline{I}} - \frac{n\overline{\overline{\alpha}}}{3\epsilon_0\varepsilon_m}) = \overline{\overline{\alpha}} \cdot \mathbf{E} + \frac{1}{2}(\overline{\overline{\delta}} \cdot \nabla) \cdot \left(\mathbf{E} + \frac{\mathbf{P}}{3\epsilon_0\varepsilon_m}\right)$$

$$-(\overline{\overline{m}}')^T \cdot \mathbf{B} - (\overline{\overline{m}}')^T \cdot n(\overline{\overline{m}}') \cdot \left(\overline{\overline{I}} + \frac{n\overline{\overline{\alpha}}}{3\epsilon_0\varepsilon_m}\right) \cdot \mathbf{E}. \tag{3.141}$$

Now, let us recall Eqs. (3.43) and (3.48). Keeping in (3.43) only zero-order and first-order terms and substituting (3.48), we have

$$\mathbf{P} = \overline{\overline{a}} \cdot \mathbf{E} + \frac{1}{2}(\overline{\overline{d}} \cdot \nabla) \cdot \mathbf{E} - \overline{\overline{g}} \cdot \mathbf{B}. \tag{3.142}$$

Comparing (3.141) and (3.142), we see that

$$\frac{\overline{\overline{a}}}{n} = (\overline{\overline{I}} - \frac{n\overline{\overline{\alpha}}}{3\epsilon_0\varepsilon_m})^{-1} \cdot \left[\overline{\overline{\alpha}} - (\overline{\overline{m}})^T \cdot n\overline{\overline{m}}' \cdot (\overline{\overline{I}} + \frac{n\overline{\overline{\alpha}}}{3\epsilon_0\varepsilon_m})\right], \tag{3.143}$$

and for the MEC parameter we have

$$\frac{\overline{\overline{g}}}{n} = -\left(\overline{\overline{I}} - \frac{n\overline{\overline{\alpha}}}{3\epsilon_0\varepsilon_m}\right)^{-1} \cdot (\overline{\overline{m}}')^T. \qquad (3.144)$$

The expression in the right-hand side of (3.144) has the first order of smallness. As to (3.143), we have to skip in this formula the second and third-order terms, that gives, finally:

$$\overline{\overline{a}} = n\left(\overline{\overline{I}} - \frac{n\overline{\overline{\alpha}}}{3\epsilon_0\varepsilon_m}\right)^{-1} \cdot \overline{\overline{\alpha}}. \qquad (3.145)$$

It is easy to prove that the medium susceptibility is the Lorentzian function. Any component of the p-dipole polarizability α_{ii} of any passive scatterer experiences the Lorentz's type resonance [53]. Neglecting for simplicity of writing the dissipative and scattering losses we may write

$$\alpha_{ii} = \frac{F}{1 - \left(\frac{\omega}{\omega_0}\right)^2}, \quad F = \text{const}(\omega). \qquad (3.146)$$

Substituting (3.146) into (3.145) we after a simplest algebra obtain

$$a_{ii} = \frac{F/\left(1 - \frac{nF}{3\varepsilon_0\varepsilon_m}\right)}{1 - \left(\frac{\omega}{\omega_1}\right)^2}, \quad \omega_1 = \omega_0\sqrt{1 - \frac{nF}{3\varepsilon_0\varepsilon_m}}. \qquad (3.147)$$

It results in the expression (1.8) for the effective permittivity with frequency independent parameter F_1. As to the MEC parameter, the analogue of (3.146) for the magnetoelectric polarizablity of a particle $m'_{ij} = j\omega\gamma_{ji}$ differs by the factor ω (see [79]):

$$m'_{ij} = \frac{\omega F'}{1 - \left(\frac{\omega}{\omega_0}\right)^2}, \quad F' = \text{const}(\omega),$$

that results in the factor ω also in $F_1(\omega)$ if we refer formula (1.8) to any components of the magneto-electric coupling tensor of the medium. Also, for all components of the MEC tensor $F_0 = 0$. The account of losses changes nothing in this proof. The Lorentzian frequency dispersion of the microscopic parameters (polarizabilities) results in the Lorentzian frequency dispersion of the macroscopic parameters (susceptibilities). In the correct model it obviously must be so.

Comparing formulas (3.140) and (3.144), we see that our result really meets the reciprocity principle (3.72), expressed on the macroscopic level by (3.137). In fact, the tensor $\overline{\overline{M}}'$ of the susceptibility of the magnetic polarization of the medium to the electric field

$$\overline{\overline{M}}' = n\overline{\overline{m}}' \cdot (\overline{\overline{I}} - \frac{n\overline{\overline{\alpha}}}{3\varepsilon_0 \varepsilon_m})^{-1} \qquad (3.148)$$

is equal to $-\overline{\overline{g}}^T$, where $\overline{\overline{g}}$ given by (3.144) is the susceptibility of the electric polarization of the medium to the magnetic field.

So, we expressed the p-dipole and m-dipole susceptibilities of the medium through the known microscopic tensors. Let us find now the quadrupole susceptibility. From formulas (3.44), (3.132), and (3.138) we have:

$$\frac{\overline{\overline{Q}}'}{n} = \overline{\overline{q}}' \cdot [\overline{\overline{I}} + \frac{n\overline{\overline{\alpha}}}{3\varepsilon_0 \varepsilon_m} \cdot (\overline{\overline{I}} - \frac{n\overline{\overline{\alpha}}}{3\varepsilon_0 \varepsilon_m})^{-1}]. \qquad (3.149)$$

In this relation we have to skip the term with n^2. Then (3.149) yields to the equation:

$$\overline{\overline{Q}}' = n\overline{\overline{q}}' \cdot (\overline{\overline{I}} - \frac{n\overline{\overline{\alpha}}}{3\varepsilon_0 \varepsilon_m})^{-1}. \qquad (3.150)$$

Parameters $\overline{\overline{\alpha}}$, $\overline{\overline{M}}'$ and $\overline{\overline{Q}}'$ enter MEs (3.70) and (3.71) as components of the permittivity, permeability, and MEC parameter, as it is clear from equivalent formulas (3.66) and (3.67). In principle, we can add also the expression for a quasi-material parameter d_{ijk}, following from (3.131) and (3.150). We call it quasi-material because this parameter does not contribute into our MEs. However, it also can be interesting for some special cases. In the index form the microscopic relation for this macroscopic parameter reads as:

$$d_{ijk} = nq'_{kil}(I_{lk} - \frac{n\alpha_{lk}}{3\varepsilon_0 \varepsilon_m})^{-1}. \qquad (3.151)$$

As to the Maxwell–Garnett model for media with the second-order SD, this model for the simple special case— media of p- and m-dipoles—has already been presented above. Its generalization to the case when the electric quadrupoles and octupoles and magnetic quadrupoles are present was done in [1]. It was shown that Eqs. (3.78) ensuring the reciprocity in the second order of smallness follow from the reciprocity of the microscopic response. As well as in the media with SD of the first order, the Maxwell–Garnett model for such multipolar media keeps the frequency dispersion of the EMPs Lorentzian if the dispersion of the multipole polarizabilities is Lorentzian.

The important results of this analysis deserve to be presented here. First, formulas (3.145), (3.148) and (3.150) keep valid if we replace

$$\frac{\overline{\overline{\alpha}}}{3\varepsilon_0\varepsilon_m} \to \frac{\overline{\overline{\alpha}}}{3\varepsilon_0\varepsilon_m} + \frac{\overline{\overline{\beta}}}{3\mu_0},$$

where β is the microscopic magnetic polarizability defined by the generalization of the relation (3.136):

$$m_i = m'_{ij} E_j^{loc} + \beta_{ij} H_j^{loc}, \qquad (3.152)$$

where the dependence on the curl of the local field in (3.45) is equivalently replaced by the dependence on the local magnetic field. Second, the m-dipole part of the magnetic susceptibility is given by

$$\overline{\overline{G}} = n(\overline{\overline{I}} - \frac{n\overline{\overline{\alpha}}}{3\varepsilon_0\varepsilon_m} - \frac{n\overline{\overline{\beta}}}{3\mu_0})^{-1} \cdot \overline{\overline{\beta}}. \qquad (3.153)$$

It is possible to show in the case when the high-order multipoles are absent (for arrays of p-and-m particles) formulas (3.145), (3.148) and (3.153) transit to earlier known relations. For example, these formulas for an array of parallel canonical CPs meet Eqs. (2.32), (2.33), (2.34). To prove it the relation for three scalar polarizabilities of a canonical CP (see [132])

$$a_{ee}^{zz} a_{mm}^{zz} + \left(a_{me}^{zz}\right)^2 = 0 \qquad (3.154)$$

was used in [1]. Relation (3.154) nullifies all the terms proportional to n^2 in (2.32), (2.33), (2.34) and these formulas become equivalent to (3.145), (3.148) and (3.153) with substitutions $\alpha_{zz} = a_{ee}^{zz}$, $\alpha_{xx,yy} = a_{ee}^t$, $m'_{zz} = a_{me}$ and $\beta_{zz} = a_{mm}$.

Notice that analogues of (3.154) for whatever solid BA particles were derived in [133]. To conclude this section, let us note that the second-order corrections to the EMPs calculated in the first-order theory, namely to formulas (3.145), (3.148) and (3.150) are (for non-resonant composites) insignificant. The main mixing rule of the second-order theory is Eq. (3.153).

3.8 WSD in a Non-Resonant Array of Dielectric Spheres

The aim of the present section is to outline the frequency bounds of the WSD on a practical example. At which frequencies the effect of WSD becomes noticeable. At which frequencies does the theory of WSD become not applicable because the medium loses the effective continuity? Media with the significant b-parameter represent a too difficult implementation of WSD in order to serve a good example. The best example is definitely the simplest one—a simple cubic lattice of spheres. This lattice does not possess the bianisotropy and the phenomenon of WSD manifests in it only in the form of artificial magnetism.

3.8.1 About Artificial Magnetism of Dielectric Spheres

It was already mentioned above that at the magnetic Mie resonance an array of densely packed spheres is effectively continuous if the spheres are sufficiently small and their refractive index is sufficiently high. As it has been already pointed out the magnetic Mie resonance is the fundamental one—it is located on the frequency axis noticeably below the electric dipole resonance.

Here, it is worth to note that researchers can be interested in the intersection of the multipolar Mie resonance bands. For example in order to engineer the simultaneously negative permittivity and permeability of the effective medium of dielectric particles (it is a reasonable approach allowing one to avoid dissipation in the metal) one may try to combine the m-dipole and p-dipole Mie resonances. There are two ways to this overlapping. The first way is to change the shape of the dielectric particle from a sphere to a spheroid. In strongly oblate spheroids (tablets) the p-dipole and m-dipole resonances intersect strongly. However, a strongly oblate spheroid unlike a sphere is not a high-quality scatterer. Even very small dielectric losses in the material of dielectric tablets for a medium composed by such tablets will result in a significant decay. Moreover, the most interesting realization of the doubly negative medium would be an isotropic one, and a tablet is not a very appropriate building block for an isotropic array.

For anisotropic doubly negative media tablets can be used but one needs a material with very high permittivity and very low imaginary part. Up to now solid media with permittivity higher than 100 (barium-lithium-titan ceramics) are lossy [186]. As to low-loss ceramics operating at microwaves (permittivity attains 90–100) or semiconductors operating in the optical range (permittivity of silicon or germanium attains 15–16 beyond the frequency regions of noticeable losses) refractive indexes of these materials are not high enough to create the negative permittivity and permeability at the same frequency.[j]

Another way to create all-dielectric doubly negative materials is to use a two-phase composite of spheres. Spheres of one smaller diameter may have the magnetic Mie resonant at the same frequency where the larger spheres have the p-dipole resonance. The m-dipoles of larger spheres are negligible in this combined p-m resonance band. As to p-dipole of smaller spheres, they are practically quasi-static in this band and can be taken into account via the enhanced permittivity of the host medium. In this way we come to an effective medium formed by p-dipole and m-dipoles resonating at the same frequency. Such an MM will be analyzed in the last chapter of this book.

In the present chapter we consider the impact of the artificial magnetism below the magnetic Mie resonance. Many researchers believe that this effect of WSD is important only in the frequency region where the effective permeability is resonant. However, we have pointed out how important is such a non-resonant manifestation of WSD as chirality of natural media. The effect is microscopically weak, but has important implications on the macroscopic level. The implications of the artificial magnetism are not as spectacular as the polarization rotation, but we will see how the account of this effect improves the accuracy of the effective-medium model.

---

[j]This text was written in 2003. Now, the situation has changed. Low-loss ferroelectric ceramics with the permittivity of the order of several hundreds and tangent of dielectric losses of the order of 10^{-4} have been created [187].

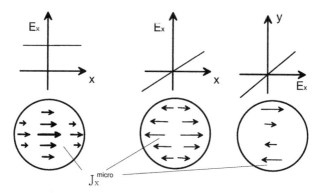

Figure 3.5 A small sphere in the uniform or linearly varying local field $E = x_0 E_x$. The non-uniformity of polarization currents J_x^{micro} in the uniform field results in the negligibly small octupole O_{xxx}. The field E_x linearly varying along x results in the negligibly small quadrupole Q_{xx}. The linear variation of E_x along y results in both off-diagonal quadrupole $Q_{xy} = Q_{yx}$ and magnetic dipole M_z whose fields exactly compensate one another.

3.8.2 A Simple Cubic Lattice of Dielectric Spheres: Theory

The improvement in the accuracy granted by the account of artificial magnetism will be shown using the boundary problem solution when the effective medium was composed of lossless dielectric spheres with permittivity $\varepsilon_s = 10$ or $\varepsilon_s = 30$.

Within the framework of the second-order theory the dipole moment of a sphere may respond to both uniform and quadratic parts of the local field coordinate dependence. The response to the linearly varying part (that changes the sign at the particle center) is evidently absent due to the particle symmetry. It is clear, for example, from Fig. 3.5. Here, for simplicity of writing in this subsection we denote the local field simply as **E**.

All multipole microscopic polarizabilities of a sphere can be found from the rigorous solution of a boundary problem. Diffraction of a plane wave by a dielectric sphere allows the scattered field to be presented as the series of spherical functions with the so-called *Mie coefficients*. These spherical functions represent the fields of the *spherical multipoles* induced in a sphere by a plane wave [53] and these Mie coefficients describe the spherical multipoles. The relations between the Cartesian and spherical multipoles is well known [53, 56] and our multipoles can be analytically found from this solution as well.

So, in principle, we could calculate all our multipole polarizabilities based on the solution of the plane-wave diffraction problem. Integer polarizabilities (referred to the local field at the center) can be found analytically and differential polarizabilities (referred to the spatial derivatives local field at the center) can be found numerically using the superposition principle for two incident waves (see above).

However, our study concerns the low-frequency region—below the magnetic Mie resonance. In this band only p-dipoles and m-dipoles can be taken into account. The account of m-dipoles is the same as the account of WSD. Looking at Fig. 3.5 it is easy to see that the uniform part of x-polarized electric field induces the polarization currents J_x^{micro} corresponding to the p-dipole mode, whereas the part of **E** linearly varying along x induces the diagonal quadrupole mode whose response to the electric field does not enter our EMPs. The part of **E** linearly varying along x induces the m-dipole moment and off-diagonal electric quadrupole moment. However, the corresponding quadrupole polarizability cancels out with that induced by the part of **E** polarized along y and linearly varying along x and does not enter our EMPs determined by Eqs. (3.89) and (3.153).

Now, let us prove that the part of $\mathbf{E} = \mathbf{x}_0 E_x$ quadratically varying along x does not contribute into its dipole moment. This contribution can be an only reason of the nonzero b-parameter of an effective isotropic medium. And we will see that for an array of spheres the b-parameter is zero.

In accordance with the concept of WSD, the p-dipole moment of a sphere impinged by a local electric field polarized along x takes form:

$$p_x = \alpha_0 E_x + \alpha_2 \frac{\partial^2 E_x}{\partial x^2}. \tag{3.155}$$

Here, α_0 and α_2 are scalars, and α_0 is the well-known quasi-static polarizability of a sphere:

$$\alpha_0 = 3\varepsilon_0 \frac{\varepsilon_s - \varepsilon_m}{\varepsilon_s + 2\varepsilon_m} V_s, \tag{3.156}$$

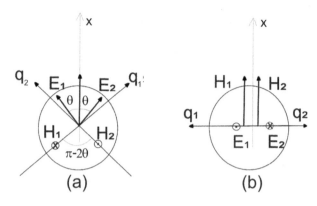

Figure 3.6 Illustration to the proof of the fact that the polarizabilities of a dielectric sphere within the framework of the second-order theory do not contain the response to the spatial co-derivatives of the electric field (a) and that the response to the spatial cross-derivatives of the electric field yields to the magnetic polarizability (b).

where V_s is the sphere volume. The coefficient α_2 entering the second term in the right-hand side of (3.155) can be found from the superposition principle.

Let our sphere be excited by two plane waves so that the resulting wave is standing along y and travelling along x, as shown in Fig. 3.6a. In this regime two wave vectors $\mathbf{q}_1$ and $\mathbf{q}_2$ of two incident waves (having the same absolute values $q_1 = q_2 = q$) intersect with a certain angle that we may denote $\pi - 2\theta$. Then the electric fields of these waves (let them have the unit amplitude $E_1 = E_2 = 1$) form the angle θ with axis x.

Two partial p-dipole moments induced by these two waves are directed along $\mathbf{E}_1$ and $\mathbf{E}_2$, respectively, and equal to α_{Mie}, where α_{Mie} is the dynamic dipole polarizability of a sphere, known from the Mie theory. In accordance with these evident observations, we have for the total dipole moment $p_y = 0$ and $p_x = 2\alpha_{\text{Mie}} \cos \theta$.

Compare this formula with (3.155) and see that $\alpha_2 = 0$ in its right-hand side. In fact, $\partial^2 E_x/\partial x^2 = -q_x^2 E_x = -q_x^2 \cos\theta$, where $q_x = q \sin\theta$, and the second term in (3.155) is equal to $2\alpha_2 q^2 \cos^2\theta \sin\theta$. Formula (3.155) gives for the induced dipole moment the result $p_x = 2\alpha_0 \cos\theta + 2\alpha_2 q^2 \cos^2\theta \sin\theta$. It is evident that this result is non-physical if the second term is nonzero. So, in (3.155) $\alpha_2 = 0$, and the dipole moment is simply proportional to the local electric field at the center

$$\mathbf{p} = \alpha_0 \mathbf{E}. \tag{3.157}$$

The theory by Mie gives a similar formula $\mathbf{p} = \alpha_{\text{Mie}} \mathbf{E}$ for a sphere, but the coefficient α_{Mie} is different from (3.156) and expressed via spherical Bessel's functions of the optical size of the sphere. However, in our frequency range—below the magnetic Mie resonance and well below the electric Mie resonance we may put $\alpha_0 = \alpha_{\text{Mie}}$ with high accuracy.

The m-dipole polarizability can be either retrieved from the Mie coefficients or found from the excitation scheme with two oppositely travelling waves illustrated by Fig. 3.6b. Here at the sphere center $H_1 = H_2 = H$ and the vector $\mathbf{H}$ is oriented along x, whereas $E_1 = -E_2$. Then the p-dipole is not induced and the whole scattered field of the sphere is that of the m-dipole, excited by the uniform magnetic field:

$$\mathbf{m} = \beta \mathbf{H}. \tag{3.158}$$

So, in the case of spheres the WSD really yields to the excitation of the m-dipole by the local magnetic field. Relations (3.157) and (3.158) means that our array is a simplest magneto-dielectric to which the usual Maxwell Garnett model—Eqs. (2.22) and (2.23)—is applicable with substitutions $a_{ee} \equiv \alpha$ and $a_{mm} \equiv \beta$. Both p-dipole and m-dipole polarizabilities can be found in books [38, 53, 55, 64].

3.8.3 A Simple Cubic Lattice of Spheres: Calculations

We can easily solve the boundary problem for a slab formed by N monolayers of dielectric sphere treating it as a magneto-dielectric medium with effective permittivity ε and permeability μ:

$$D_i = \varepsilon_0 \varepsilon E_i, \quad B_i = \mu_0 \mu H_i$$

whose top and bottom interfaces coincide with the effective boundaries of the simple cubic lattice, i.e., the slab thickness is equal Nd. The power reflection coefficient R of the effective slab is given by the standard Fresnel-Airy formula:

$$R = \frac{|R_\infty(1 - e^{2ik_0\sqrt{\varepsilon\mu}D})|^2}{|1 - R_\infty^2 e^{2ik_0\sqrt{\varepsilon\mu}D}|^2}, \quad R_\infty = \frac{\sqrt{\mu/\varepsilon} - 1}{\sqrt{\mu/\varepsilon} + 1} \tag{3.159}$$

that represents the square power of the absolute value of the amplitude reflection coefficient for the normal incidence. A similar formula is known also for the oblique incidence (see [53]). The spheres with $\varepsilon_s = 10$ are assumed to be located in free space and their diameter a is twice smaller than the period d. The result of this calculation was compared with that of the quasi-static model neglecting the magnetic moments of the spheres—letting $\mu = 1$ in (3.159). It was also compared with full-wave calculations using the so-called GMM code suitable for arbitrary arrays of dielectric spheres. This code was described in [188] and is available in the Web (https://code.google.com/archive/p/scatterlib/).

In Fig. 3.7a,b, we can see the reflection coefficient R of two arrays, one has $N = 4$ monolayers and another has $N = 20$. R is presented as a function of the normalized frequency (optical size of the unit cell) qd. Here by definition $q = k_0\sqrt{\varepsilon\mu}$ is the wave number of the effective medium. In our calculations values ε and μ were real. In Fig. 3.7a we show $R(qd)$ for $N = 4$ and in Fig. 3.7b we show it for $N = 20$.

The WSD (the account of the m-dipoles induced in our spheres) becomes relevant starting from $qd = 0.5$–0.6 where the accuracy of the quasi-static model worsens. The homogenization model taking into account the WSD keeps quite accurate in the case when $N = 4$ until $qd \approx 1$ (the magnetic Mie resonance holds when $kd \approx 2.2$). In the case $N = 20$ the WSD is also qualitatively adequate until $qd = 1$. However, quantitatively it becomes inaccurate (predicts minima of the reflectance when the exact solution predicts maxima and vice versa) for $qd > 0.3$. This is so because the effective-medium model even taking into account the fine effects of WSD is hardly applicable at frequencies of the second and higher Fabry–Perot resonances. For $N = 20$ the first Fabry–Perot resonance holds at the normalized frequency $qd \approx 0.15$ and the second—at $qd \approx 0.3$. At these resonances even a slight error in the effective permittivity or permeability (whose product gives the square power of the refractive index) results in the significant error for the reflectance and transmittance. The thicker is the slab the higher is this error. It is not surprising because the Fabry–Perot resonance is an interference phenomenon. It predicts the maximum of the interference when the phase shift between interfering waves is equal 0, 2π, $4\pi \cdots$ and

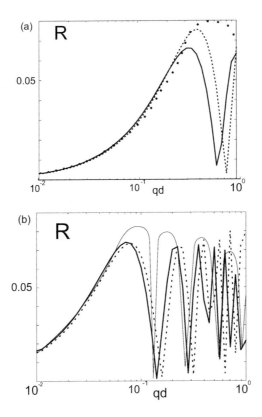

Figure 3.7 Reflectance from a finite-thickness simple cubic lattice of dielectric spheres with $\varepsilon_s = 10$. The reflectance for the cases of 4 layers (a) and 20 layers (b) is shown versus the normalized frequency qd. Thick solid line corresponds to the Maxwell Garnett effective medium model taking into account the m-dipoles. Dashed line corresponds to the full wave simulations. Dotted line in (a) and thin solid line in (b) correspond to the Maxwell Garnett model ignoring the m-dipoles of spheres.

minimum if this phase shift is equal π, 3π, 5π $\cdots$. Let the error in the array refractive index $n = \sqrt{\varepsilon\mu}$ brought by the effective-medium model be equal δn. Then the error in the effective medium wave number is $\delta q = k_0 \delta n$. It is evident that the error in the phase shift of the wave transmitted through the slab is equal $k_0 \delta n N d$, i.e., is proportional to N.

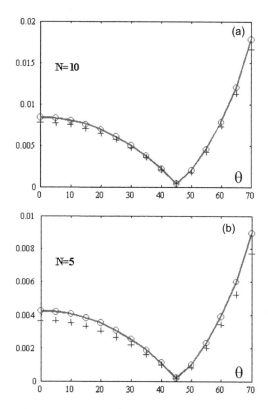

Figure 3.8 Reflectance from a 10-layered (a) and 5-layered (b) simple cubic lattice of spheres with $\varepsilon_s = 30$ (**E** is polarized in the incidence plane) versus incidence angle $\theta°$ (normalized frequencies are $qd = 0.1$ and $d = 1.5a$). Circles correspond to the account of m-dipoles, crosses correspond to the model neglecting them, solid lines—full-wave simulations.

So, the frequency bound where the homogenized model of the array loses the practical accuracy depends on the thickness of the medium sample. However, this frequency bound—$qd = 1$ for $N = 4$ and $qd = 0.3$ for $N = 20$—is related with the inaccuracy of the predicted phase shift and not with the inadequacy of the model as such. It is not the frequency bound at which the effective medium becomes a photonic crystal. As to the qualitative accuracy of the model—adequate shape of the curve, the situation is more favorable for our model. For the case $N = 4$ the account of WSD (m-dipoles) improves this accuracy at $qd > 0.5$. For the case $N = 20$ the account of WSD improves the qualitative accuracy at $0.1 < qd < 0.9$.

In order to judge on the applicability of the effective-medium model for the oblique incidence corresponding calculations were done for the low frequency region—below the fist Fabry–Perot resonance. For $N > 3$ the accuracy granted by our the model in these calculations is excellent. In Fig. 3.8 we show the reflectance of a TM-polarized wave versus the incidence angle $\theta°$ for the case $d = 1.5a$, at normalized frequency $qd = 0.1$ for ceramic spheres with $\varepsilon_s = 30$. The number of layers is equal either $N = 10$ (a) or $N = 5$ (b). Here, the nice correspondence of all three methods shows that the effective-medium model in this frequency range is fully adequate for all incidence angles and that the impact of WSD in this case is also weak. Moreover, m-dipoles in this case give more contribution for the normal incidence and have no impact in the vicinity of the Brewster angle.

In general, the inspection of the oblique incidence does not change the frequency bound until which the effective medium model allows the accurate calculation of the reflectance. In what concerns the numerical accuracy, this bound decreases with the thicknesses of the effective medium slab. As to qualitative accuracy, in the first example ($\varepsilon_s = 30$) this bound is $(qd) = 0.9$–1.0, well below the Mie resonances. The account of WSD noticeably improves the accuracy of the homogenization model even in this range.

3.9 Conclusions of This Chapter

We have derived MEs for composite or molecular media with WSD and built the Maxwell Garnett model relating the EMPs with the response of the individual particle to the local field. Our Maxwell Garnett model is based on CMLL formulas which were at this stage simply postulated.

As a rule, WSD yields to two phenomena—artificial magnetism and bianisotropy. In both these cases, the usual Maxwell boundary conditions allow one to solve boundary problems in the frequency regions free of the resonances of the constitutive elements. With an example of the effective layer of spheres we have shown that even at low frequencies the account of the artificial magnetism offers a noticeably higher accuracy of the solution. However, the homogenization model as such is an approximation and comprises an error. The medium model is more accurate in the solution of the boundary problem if the composite layer is sufficiently thin. For thick samples the error of the homogenization model accumulates in the phases of internally reflected waves and the frequencies of the Fabry–Perot resonances are predicted wrongly.

Besides media with artificial magnetism and bianisotropy, the second-order spatial dispersion can arise in media with an additional EMP we called the b-parameter. An example of this medium is an optically dense though geometrically dilute composite of thin metal needles. In this case the MEs obviously include the spatial co-derivatives of the mean electric field. Co-derivatives of **E** cannot be reduced to the magnetic field or removed by the redefinition of **H** and **D**. In this case the main question is that of boundary conditions. Maxwell's boundary conditions are not sufficient for such media, and additional boundary conditions are needed in spite of the effective continuity of the medium. Otherwise the boundary problem cannot be solved.

However, even for composite media without the b-parameter Maxwell's boundary conditions in order to be applied demand of us to determine the position of the effective boundary of our array. We definitely have no right to postulate this boundary as it was done by Madelung, Ewald and Oseen—as the plane centered by the top particles of the lattice. Our theory is essentially based on the account of the finite particle size. In the next chapter we will see where the effective boundary is located and why the homogenization model allows us to satisfy Maxwell's boundary conditions without involving the extinction principle. Of course, the extinction principle still holds for our composites. Definitely, the mission of the interface particles to cancel the incident wave inside the array keeps [114] and of course the CMLL formulas fail for the interface particles. However, the homogenization of the medium interface based on the ideas of P. P. Drude does not need to involve the CMLL formulas.

Here, we will also concern the case when the insight of the bulk material for a composite layer becomes inadequate, namely, when the layer comprises only 1–2 constitutive particles across it. In fact, the main role of the interface particles is to cancel the incident wave inside the array. However, if there are only two monolayers in the array there is no bulk in which the incident wave could be compensated. The extinction principle becomes meaningless, whereas the CMLL formulas fail for both monolayers by an evident reason. In this situation the revised theory of Drude's transition layers helps to solve the boundary problem with the high accuracy.

Here, it is reasonable to mention the so-called *surface states* we will not concern in this book. Surface states can be treated as manifestations of SD *for the surface* of a composite sample. In fact, we have considered one of these effects—excitation of polaritons [108, 111]. This effect in order to be noticeable does not demand the resonance of the constitutive particles. In the combination with these resonances the wave retardation per unit cell may result in the surface states [66]. The most known example of such surface states is the so-called *Wood anomaly*—a surface wave excited by an obliquely incident TM-polarized plane wave in arrays of resonant particles [189].

Chapter 4

Revision of the CMLL Formulas and Boundary Conditions for Thin Composite Layers

4.1 Retardation Effects in the CMLL Formulas

4.1.1 Interaction of Non-Resonant Particles

Here we start to derive the CMLL formulas (3.132), (3.133) in order to evaluate the bounds of their applicability and, perhaps, to deduce the correction terms. The key assumption of the derivation is that $kd < 1$ is the parameter of smallness. As above $k = k_0\sqrt{\varepsilon_m}$ is the wave number of the matrix, and the effective medium wave number is denoted as q. Here we deal so much with true fields that it is reasonable to omit the index *true* for both true (microscopic) fields and polarizations. In this section all averaged values will be denoted by brackets <>.

Local field acting on a i-th particle referred to the center of the i-th elementary cell (radius-vector $\mathbf{R}_i$) in the case when the array comprises N particles can be presented as follows:

$$\mathbf{E}_i^{\text{loc}} = \mathbf{E}^{\text{ext}}(\mathbf{R}_i) + \sum_{j \neq i} \mathbf{E}_j^{\text{part}}(\mathbf{R}_i), \qquad (4.1)$$

Composite Media with Weak Spatial Dispersion
Constantin Simovski
Copyright © 2018 Pan Stanford Publishing Pte. Ltd.
ISBN 978-981-4774-83-3 (Hardcover), 978-1-351-16624-9 (eBook)
www.panstanford.com

where $\mathbf{E}^{\text{ext}}$ is the field of external sources refracted to the matrix and $\mathbf{E}_j^{\text{part}}$ is the field produced by j-th polarized particle of the array. Microscopic field in the arbitrary point shifted by a small vector $\mathbf{r}$ from the center of the i-th unit cell can be written in a form:

$$\mathbf{E}(\mathbf{R}_i + \mathbf{r}) = \mathbf{E}^{\text{ext}}(\mathbf{R}_i + \mathbf{r}) + \sum_{j \neq i} \mathbf{E}_j^{\text{part}}(\mathbf{r} + \mathbf{R}_i) + \mathbf{E}_i^{\text{part}}(\mathbf{R}_i + \mathbf{r}). \quad (4.2)$$

The notations are illustrated by Fig. 4.1.

The mean field and mean electric polarization at the arbitrary point $\mathbf{R}$ inside the array (up to its matrix surface) are defined as:

$$<\mathbf{E}>(\mathbf{R}) = \frac{1}{V_{av}} \int_{V_{av}} \mathbf{E}(\mathbf{R} + \mathbf{r}') d^3\mathbf{r}', \quad (4.3)$$

$$<\mathbf{P}>(\mathbf{R}) = \frac{1}{V_{av}} \int_{V_{av}} \mathbf{P}(\mathbf{R} + \mathbf{r}') d^3\mathbf{r}', \quad (4.4)$$

respectively. Here V_{av} is the averaging volume, which is supposed (following to the modern understanding of Ewald's averaging procedure) to be equal to the cell volume $V = d^3$. The volume of averaging is centered at the observation point. Here we adopt the quasi-static averaging algorithm, as it is always done in classical theories of homogenization. We have discussed this concept above, and below we will revise it. However, for the instance, we use formulas (4.4) as definitions of the averaging. Microscopic polarization is absent outside particles and for the center of i-th particle we have $<\mathbf{P}>(\mathbf{R}_i) = \mathbf{p}_i/V_{av}$.

For the mean field at the center of the i-th particle equation (4.2) reads as follows:

$$<\mathbf{E}>(\mathbf{R}_i) = \frac{1}{V_{av}} \int_{V_{av}} \mathbf{E}^{\text{ext}}(\mathbf{R}_i + \mathbf{r}) dV + \frac{1}{V_{av}} \sum_{j \neq i} \int_{V_{av}} \mathbf{E}_j^{\text{part}}(\mathbf{r} + \mathbf{R}_i) dV$$

$$+ \frac{1}{V_{av}} \int_{V_{av}} \mathbf{E}_i^{\text{part}}(\mathbf{R}_i + \mathbf{r}) dV. \quad (4.5)$$

We consider the structure which is nearly a simple cubic lattice (perhaps with small random deviations in the positions of particles). We do not impose limitations on the total amount of particles, assuming that it is reasonably large ($N \gg 1$), and also we did not

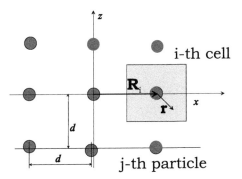

Figure 4.1 Illustration to the calculation of the interaction field at the center of i-th unit cell ($R = R_i$, $r = 0$) and its averaged value at the same point.

restrict the shape of the sample. If we deduce the CMLL with small correction terms for this case and establish the bounds of validity for these formulas, we will be sure that the anisotropic variants of CMLL formulas also hold. The assumption of a generic simple cubic lattice (i.e., cubic unit cell) strongly simplifies our derivations.

Assume that the external sources are distanced from the array sufficiently so that their field can be approximated by a plane wave propagating, for example, along z. Then the external field is nearly homogeneous over the unit cell, and its variation due to the phase shift is small. Strictly speaking, it is not negligible because we assume that our theory is valid up to the frequency at which $kd = 1$. For example, if the external field is that of a plane wave propagating along z in order to be strict we have to write:

$$<\mathbf{E}^{ext}>(\mathbf{R}_i) = \frac{\mathbf{E}_{ext}(\mathbf{R}_i)}{V_{av}} \int_{-d/2}^{d/2} e^{-jkz}\,dxdydz = \mathbf{E}_{ext}(\mathbf{R}_i)\frac{2\sin\left(\frac{kd}{2}\right)}{kd}.$$

In the second order of smallness $2(\sin(kd/2)/kd) \approx 1 - (kd)^2/6$ and $<\mathbf{E}^{ext}>(\mathbf{R}_i) \approx \mathbf{E}^{ext}>(\mathbf{R}_i)[1 - (kd)^2/6]$. However, as we see below, this complexity will bring nothing but an error to the result. Therefore, it is reasonable to put in our derivations $\sin(kd/2) \approx (kd/2)$ that gives

$$<\mathbf{E}^{ext}>(\mathbf{R}_i) \approx \mathbf{E}_{ext}(\mathbf{R}_i). \tag{4.6}$$

Subtracting (4.1) from (4.5) we have, taking (4.6) into account

$$<E>(R_i) - E_i^{loc} = \sum_{j \neq i} \left[<E_j^{part}>(R_i) - E_j(R_i) \right] + <E_i^{part}>(R_i). \tag{4.7}$$

Let us expand $<E_j^{part}>$ into the Taylor series in the vicinity of the point R_i:

$$<E_j^{part}>|_{r=0} = \frac{1}{V_{av}} \int_{V_{av}} [E_j^{part}|_{r=0} + r_\alpha \nabla_\alpha E_j^{part}|_{r=0}$$
$$+ \frac{1}{2} r_\alpha r_\beta \nabla_\alpha \nabla_\beta E_j^{part}|_{r=0} + \frac{1}{6} r_\alpha r_\beta r_\gamma \nabla_\alpha \nabla_\beta \nabla_\gamma E_j^{part}|_{r=0}$$
$$+ \frac{1}{24} r_\alpha r_\beta r_\gamma r_\delta \nabla_\alpha \nabla_\beta \nabla_\gamma \nabla_\delta E_j^{part}|_{r=0} + \ldots] dV. \tag{4.8}$$

Here Greek letters denote Cartesian coordinates. The integrand in (4.8) is a series whose second and fourth (and all even terms) give no contribution into the integral being odd functions of the coordinates. Third and fifth terms give the zero contribution either if $\alpha \neq \beta$ or if $\gamma \neq \delta$. As a result, we have:

$$<E_j^{part}>|_{r=0} - E_j^{part}|_{r=0} = \frac{1}{2V_{av}} \nabla_\alpha^2 E_j^{part}|_{r=0} \int_{V_{av}} r_\alpha^2 dV$$
$$+ \frac{1}{24V_{av}} \nabla_\alpha^2 \nabla_\beta^2 E_j^{part}|_{r=0} \int_{V_{av}} r_\alpha^2 r_\beta^2 dV + \ldots. \tag{4.9}$$

First, let us neglect in (4.9) all terms of the order $(kd)^4/720$ (second term in the right-hand side of (4.9)) and higher. The signs of the omitted terms in the series are alternating, and the series is converging fast when $(kd) \leq 1$. Noticing that the field of j-th particle in the i-th cell obeys to the Helmholtz equation:

$$\Delta E_j^{part}(r) = -k^2 E_j^{part}(r), \tag{4.10}$$

and that for any α in (4.9) we have the same tabulated integral

$$\int_{V_{av}} r_\alpha^2 dV = d^5/12$$

it is easy to calculate the first term in the right-hand side of (4.9). Since for any α in the right-hand side of (4.9) we have a common factor $d^5/12$ the first term contains the Laplace operator $\sum_\alpha \nabla_\alpha^2 \equiv \Delta$. Therefore, we obtain for the value $<\mathbf{E}>_j^{part}(\mathbf{r}=0) - \mathbf{E}_j^{part}(\mathbf{r}=0)$, denoted below as $\delta \mathbf{E}_j^{part}$ the following equations:

$$\delta \mathbf{E}_j^{part} \equiv <\mathbf{E}_j^{part}>(\mathbf{r}=0) - \mathbf{E}_j^{part}(\mathbf{r}=0) = -\frac{(kd)^2}{24}\mathbf{E}_j^{part}(\mathbf{r}=0). \quad (4.11)$$

Substituting (4.11) into (4.7), we obtain:

$$<\mathbf{E}>(\mathbf{R}_i) - \mathbf{E}_i^{loc} = <\mathbf{E}_i^{part}> + \sum_{j\neq i}\delta \mathbf{E}_j^{part} = <\mathbf{E}_i^{part}> - \frac{(kd)^2}{24}\mathbf{E}^{int}. \quad (4.12)$$

In Eq. (4.12) all values in the right-hand side are taken at $\mathbf{R}_i$, and $\mathbf{E}^{int}$ denotes the interaction field—the field of all j–th particles acting on the reference i-th particle:

$$\mathbf{E}^{int} = \sum_{j\neq i}\mathbf{E}_j^{part}.$$

Equation (4.12) comprises the second-order term $(kd)^2$, which expresses the retardation effect over the unit cell. But for this retardation, the contributions of all particles surrounding the reference one into the local field would cancel out with their contributions into the mean field. Then the difference of the local and averaged field in the CMLL formula would be equal to the averaged field of the reference particle calculated at its center.

This result keeps valid also for slightly irregular bulk arrays, if the unit cell comprises one particle. Recall that our MEs are covariant and multipoles of any particle if they are not negligible can be attributed to the nodes of a cubic lattice. The only condition restricting the randomness is that the particles should not cross the bounds of the unit cell forming clusters. Our derivation shows that the difference between the local and mean fields is determined *not by the surround of the reference particle* (as it was written in [74] and repeated in many later books) but by the unit cell for which this difference is calculated. The only factor corresponding to the surround is the second-order term $(kd)^2$ in (4.12). However, we will see below that this term also should be neglected.

4.1.2 Averaging of the Reference Particle Field

The term $<\mathbf{E}^{\text{part}}>(\mathbf{R}_i)$ in (4.7) can be written in a form:

$$<\mathbf{E}^{\text{part}}>(\mathbf{R}_i) = \frac{1}{V_{av}} \int_{V_{av}} \mathbf{E}_i^{\text{part}}(\mathbf{R}_i + \mathbf{r})dV. \qquad (4.13)$$

Let us present $\mathbf{E}_i^{\text{part}}$ through the vector-potential $\mathbf{A}_i$ of this partial field and its scalar potential Φ_i:

$$\mathbf{E}_i^{\text{part}}(\mathbf{R}_i) = j\omega \mathbf{A}_i(\mathbf{R}_i) - \nabla \Phi_i, \qquad (4.14)$$

where for the vector-potential we have:

$$\mathbf{A}_i(\mathbf{R}_i + \mathbf{r}) = \frac{j\omega\mu_0 \mathbf{p}_i e^{-jkr}}{4\pi r} + \ldots \qquad (4.15)$$

Here the contributions of high-order multipoles are not shown (and recall that $k \equiv \omega\sqrt{\varepsilon_0\mu_0\varepsilon_m}$). As to the scalar potential, it is expressed as the divergence of the vector-potential (the commonly known Lorentz gauge). Let us start from the case $(kd) \ll 1$, when one can neglect both the vector-potential and the retardation in the scalar potential, assuming that $\exp(-jkr) = 1$.

4.1.2.1 Static case

In this static limit we may write the static multipole expansion for the field of the reference particle (see [38, 53, 55]), and write

$$\mathbf{E}_i^{\text{part}}(\mathbf{R}_i + \mathbf{r}) = -\frac{1}{4\pi\varepsilon_0\varepsilon_m} \nabla \left(\frac{p_\alpha r_\alpha}{r^3} + \frac{q_{\alpha\beta} r_\alpha r_\beta}{2r^5} + \frac{o_{\alpha\beta\gamma} r_\alpha r_\beta r_\gamma}{6r^7} + \ldots \right),$$

where p_α, $q_{\alpha\beta}$, $o_{\alpha\beta\gamma}$ are p-dipole, quadrupole and octupole moments. Consider the contribution of the p-dipole. Integrating the α-th component of $\mathbf{E}_i^{\text{part}} = \mathbf{E}_i^{\text{dip}}$ in (4.13) along α-th Cartesian axis we come to the surface integral:

$$<E_{i\alpha}^{\text{dip}}>(\mathbf{R}_i) = -\frac{1}{4(d/2)^3} \frac{1}{4\pi\varepsilon_0\varepsilon_m}$$

$$\int_S \left(\frac{p_\beta r_\beta}{r_s^3} + \frac{p_\gamma r_\gamma}{r_s^3} + \frac{p_\alpha d/2}{r_0^3} \right) dr_\gamma dr_\beta, \qquad (4.16)$$

where it is denoted $r_s = \sqrt{r_\gamma^2 + r_\beta^2 + (d/2)^2}$ and $S = d \times d$ is the area of two unit cell walls orthogonal to the α-th axis. It is easy to see that at the opposite walls the integrand have the same absolute value.

Two first terms in the prances in the right-hand side of (4.16) do not give a contribution into the integral since they have opposite signs at two walls orthogonal to the α-th axis. Only p_α contributes. So, each component of $< \mathbf{E}_i^{\text{part}} >$ in statics is the product of the inverse unit cell size $(1/d)$ by the drop of potential created by the reference particle on two opposite walls of the reference unit cell. Since the walls of the unit cells are not equipotential surfaces, this voltage is averaged over the surface of the wall.

We obtain, finally

$$< E_{i\alpha}^{\text{dip}} > (\mathbf{R}_i) = -\frac{p_\alpha}{3\varepsilon_0 \varepsilon_m} \frac{1}{d^3}. \quad (4.17)$$

Since $\mathbf{p} = \mathbf{P}(\mathbf{R}_i) d^3$, we have

$$< \mathbf{E}_i^{\text{dip}} > (\mathbf{R}_i) = -\frac{<\mathbf{P}>(\mathbf{R}_i)}{3\varepsilon_0 \varepsilon_m}. \quad (4.18)$$

If we neglect the second-order term in (4.12) and identify $\mathbf{E}_i^{\text{part}}$ with $\mathbf{E}_i^{\text{dip}}$ as we did above, the result (4.18) implies the CMLL formula (2.4). We saw that this result does not depend on the particle shape and holds for a cubic lattice. Similarly, it holds for a slightly irregular array.

The contribution of the quadrupole is found similarly, and it is easy to see that in statics it vanishes exactly. The contribution of the static octupole moment is equal to

$$< E_{i\alpha}^{\text{oct}} > = -\frac{< O_{\alpha\alpha\alpha} >}{72\varepsilon_0 \varepsilon_m d^2}. \quad (4.19)$$

Comparing (4.18) with (4.19) and taking into account that the octupole moment in the non-resonant case is the value of the order pa^2, we see that the contribution of the octupole moment into the difference between the local and mean fields is smaller than that of the dipole by 2 orders of magnitude. It is not so only at the octupole resonance. However, as we have discussed, the octupole resonance holds beyond the frequency region where the medium can be effectively continuous. So, we have derived CMLL formulas for statics, and our derivation is valid for any isotropic array of a finite number N particles if these particles are distributed uniformly—do not form clusters. This derivation can be easily generalized to orthorhombic lattices when the scalar factor $1/3$ will be replaced by the tensor $\overline{\overline{N}}$.

4.1.2.2 Dynamic case

Now, let us consider the dynamic case, when the parameter (kd) keeping small is comparable with unity. Then in Eq. (4.14) we cannot neglect the vector-potential. However, even in the dynamic case the contribution of high-order multipoles is negligible. The magnetic dipole does not give a contribution into the averaged field $<\mathbf{E}_i^{part}>$ because the electric field of a magnetic dipole is azimuthal with respect to its axis, i.e., is an odd function of the Cartesian coordinates. The same refers to the contribution of the particle electric quadrupole into $<\mathbf{E}_i^{part}>$.

The contribution of high-order multipoles is essential only at their resonance frequencies. In the frequency region of the medium effective continuity it is negligible, and the electric field of the reference particle averaged over its unit cell is determined only by its p-dipole.

Substituting (4.18) and (4.15) into (4.14) and assuming that $\exp(-jkr) \approx 1 - jkr$ in (4.15) we obtain

$$<\mathbf{E}_i^{part}> = -\frac{<\mathbf{P}>}{3\varepsilon_0 \varepsilon_m} + j\frac{(kd)^3 <\mathbf{P}>}{6\varepsilon_0 \varepsilon_m} - \frac{k^2 <\mathbf{P}>}{4\pi \varepsilon_0 \varepsilon_m} \int_{V_{av}} \frac{dV}{r}.$$

The last integral is not tabulated and was evaluated numerically, which gives us the result:

$$<\mathbf{E}_i^{part}> = -\frac{\xi <\mathbf{P}>}{3\varepsilon_0 \varepsilon_m} + j\frac{(kd)^3 <\mathbf{P}>}{6\varepsilon_0 \varepsilon_m}, \qquad (4.20)$$

where it is denoted $\xi = 1 - \gamma(kd)^2$, $\gamma \approx 0.4951$. Relation (4.20) is self-consistent with Eq. (4.12). Here got a correction term of the order $(kd)^3$ because put $\exp(-jkr) = 1 - jkr$. Should we take it into account or omit it?

It can be thought that the term in (4.20) proportional to $(kd)^3$ is not important for us, since the theory of WSD restricts by the accuracy in the second order of smallness. Recall that k is the wave number in the matrix, whereas q is the wave number of the effective medium. One may think that if $(kd)^3 \ll (qd)^3$, i.e., the wave shortens in the effective medium the term $(kd)^3$ is negligibly small. However, it is not so. The theory of WSD neglects the effects of the order of $(qd)^3$ in the multipole expansions because they do not result in any experimentally revealed or theoretically predicted physical effects.

However, the difference between the local and mean fields presented as the power series of (kd) refers to the substitution of the mean field in place of the local field in the Maxwell Garnett model. It is a different story. Here we must take into account the correction terms which may improve the accuracy of that substitution.

The correction term of the order $(kd)^3$ is even more important than the term $(kd)^2$ (the difference of ξ from unity). It is responsible for a basic property of the regular and slightly irregular effective media, namely for the absence of scattering losses in them. This implies the absence of scattering by every particle of the effective medium. It is the prerequisite of the medium transparency in the case of low dissipative losses. In order words, the presence of this term in the averaged field of i-th particle allows our model to be consistent with the condition of the medium transparency.

This correction term is needed even if it is numerically small. In our example with an array of dielectric spheres above it was very small since we have considered the range $(qd) \leq 1$. However, even in that example this term will be not negligible if we calculate both power reflectance R and transmittance T. If we do it as we did earlier—using formulas (2.22) and (2.23) and dynamic (Mie) electric and magnetic polarizabilities—we will see that the sum $R+T$ is not equal unity (as it must be in accordance with the power balance). This discrepancy could be treated as an optical absorption or scattering loss $A = 1 - R - T > 0$ if our array was lossy or irregular. However, for our regular and lossless array both these interpretations would be fully non-physical. In our calculations based on the Maxwell Garnett model this error is as small as 1–3% but it is not a computational error but a drawback of the non-corrected CMLL formula. Of course, in the full-wave calculations $A = 0$.

So, the classical homogenization model ignoring the term $(kd)^3$ predicts absorption or scattering losses for the structures which cannot absorb or scatter. If we make in Eqs. (2.22) and (2.23) a substitution

$$\frac{1}{3} \to \frac{1}{3} + \frac{jk^3 V_{av}}{6}, \qquad (4.21)$$

taking into account that correction term, we get rid of the discrepancy and obtain $A = 0$.

4.1.3 Modified CMLL Formulas

If we neglect the factor $(kd)^2/24$ in (4.12) this equation and Eq. (4.20) evidently result in the following generalization of the CMM formula:

$$\mathbf{E}^{loc} - <\mathbf{E}> = \frac{\xi}{3\varepsilon_0\varepsilon_m} <\mathbf{P}> + j\frac{(k^3 V_{av}/2)}{3\varepsilon_0\varepsilon_m} <\mathbf{P}>. \quad (4.22)$$

Here $V_{av} = d^3$. If the effective medium is formed by an anisotropic lattice and $V_{av} = d_x d_y d_z$, scalar factor $f = 1/3$ in the right-hand side of (4.22) must be replaced by the tensor $\overline{\overline{f}}$ of the unit cell depolarization.

If the medium is formed by p- and m-dipoles, the similar formula holds for the magnetic local and mean fields:

$$\mathbf{B}^{loc} - <\mathbf{B}> = \frac{\xi}{3\mu_0} <\mathbf{M}> + j\frac{(k^3 V_{av}/2)}{3\mu_0} <\mathbf{M}>. \quad (4.23)$$

We have already known that in this case there are no coderivatives in the relations between the electric polarizability and the local field, i.e., $\mathbf{p} = \mathbf{P}V = \alpha\mathbf{E}^{loc}$. Also, in this case

$$<\mathbf{P}> = \varepsilon_0 \kappa <\mathbf{E}> = \varepsilon_0(\varepsilon - \varepsilon_m)<\mathbf{E}>, \quad (4.24)$$

where κ is the electric susceptibility of the composite medium, and ε is its effective permittivity. Therefore, from (4.22) it follows:

$$<\mathbf{P}> V\left(\frac{1}{\alpha} - \frac{\xi}{3V\varepsilon_0\varepsilon_m} - j\frac{k^3}{6\varepsilon_0\varepsilon_m}\right) = (\varepsilon - \varepsilon_m)<\mathbf{P}>. \quad (4.25)$$

From the so-called optical theorem [107] it follows that the value inverse to the p-dipole polarizability $(1/\alpha)$ has the imaginary part even if the dipole is lossless. This imaginary part describes the so-called *radiation losses* of a dipole scatterer [107]. Adopting the time dependence $\exp(j\omega t)$ the radiative part of the inverse polarizability is equal to

$$\left(\frac{1}{\alpha}\right)_{rad} = j\frac{k^3}{6\varepsilon_0\varepsilon_m}. \quad (4.26)$$

A very simple proof of (4.26) based on the power conservation for a p-dipole scatterer can be found in [190]. We see that substituting (4.26) and (4.24) into (4.25) we obtain the effective permittivity in which the scattering losses are cancelled:

$$\varepsilon = \varepsilon_m + \left[\left(\frac{1}{\alpha}\right)_{dis} - \frac{\xi}{3V\varepsilon_0\varepsilon_m}\right]^{-1}. \quad (4.27)$$

In this equation only the dissipative part of the inverse polarizability survives:
$$\left(\frac{1}{\alpha}\right)_{dis} = -\frac{\alpha''}{|\alpha|^2},$$
where α'' is the dissipative factor in the following representation of the p-dipole polarizability of a particle: $\alpha = \alpha' - j\alpha_{rad} - j\alpha''$. Here α', α_{rad} and α'' are real values. Now we understand why the substitutions (4.21) restore the energy balance in the boundary problem with the array of spheres. In fact, relations (4.21) for the revised Maxwell Garnett algorithm can be even improved taking into account the term of the order $(kd)^2$ in $\xi = 1 - \gamma(kd)^2$:
$$\frac{1}{3} \to \frac{1-\gamma(kd)^2}{3} + \frac{jk^3 V_{av}}{6}. \qquad (4.28)$$

If the dissipative factor α'' is negligible, the effective permittivity is real and the plane wave propagates without losses. This is so for regular and slightly irregular arrays. They are transparent if the dissipative losses are low. Fully random media of particles (amorphous composites, amorphous natural materials, and fluctuating gases) comprise stochastic clusters. If we try to split such an array of particles onto unit cells some particles will obviously cross the walls of these unit cells. For such materials, the CMLL formulas hold only in the static limit. Of course, the concept of the static limit can be attributed to time-varying fields, but only at frequencies where these clusters are also optically small. For example, a foam used in microwave experiments is a composite of microscopic voids in a dielectric matrix. Voids can form clusters and these clusters are still sub-millimeter inclusions. At microwaves the static limit is applicable to a foam formed by clusters and it can be described by the effective permittivity.

However, in the optical range the CMLL formulas for the foam do not hold, since the clusters of voids are optically large in order to consider them as constitutive particles. If we consider a separate void as a constitutive particle, we notice that their radiative losses are not compensated due to the amorphous randomness. The higher is the frequency the larger is the error. Therefore, the visible light in the foam experiences a strong decay, resulting from the strong scattering of light called the *Rayleigh scattering*. The energy per unit path of the wave is lost due to the excitation of waves in the backward and lateral directions. That is why a foam is not transparent in optics and cannot be described by an effective permittivity. Amorphous media as a rule are scattering media—if the sample is optically thick it is not transparent.

Of course, the present proof of the cancellation of the radiative losses for a regular or slightly random array of particles where the p-dipoles dominate is approximate. The strict proof for regular arrays of dipole particles can be found in [190] targeted to the calculation of the effective refractive index for an orthorhombic array of p-dipoles.

If we take into account the factor $(kd)^2/24$, Eqs. (4.12) and (4.20) result in the following generalization of the CMLL formula:

$$\mathbf{E}^{loc} - <\mathbf{E}> = \frac{1 - \gamma(kd)^2 - 0.5jk^3V}{3\varepsilon_0\varepsilon_m} <\mathbf{P}> + \frac{(kd)^2}{24}\mathbf{E}^{int}. \quad (4.29)$$

Here the interaction field $\mathbf{E}^{int}$ is that produced by all other particles at the reference particle multipole center. It is the difference between the local field $\mathbf{E}^{loc}$ and that of external sources $\mathbf{E}^{ext}$. Denoting $\psi = 1 - \gamma(kd)^2 - 0.5j(kd)^3$, from (4.29) we have

$$\mathbf{E}^{loc}\left(1 - \frac{(kd)^2}{24}\right) - <\mathbf{E}> = \frac{\psi}{3\varepsilon_0\varepsilon_m}<\mathbf{P}> - \frac{(kd)^2}{24}\mathbf{E}^{ext}. \quad (4.30)$$

Eq. (4.30) is not good for us—our homogenization model implies that the relation between the local and the mean fields is unambiguous. Otherwise, we cannot obtain a set of local EMPs—each of them will be dependent on the parameters of the external field. Of course, (4.30) also allows us to express the local field through the mean field and relates somehow the microscopic and macroscopic parameters of the composite medium. However, this relationship is dependent on the external sources, which makes impossible the description of the composite in terms of its own properties. These properties will be dependent on the illumination. Formula (4.22) is an unambiguous relation which does not violate the Kramers–Kronig equations. The factor $(kd)^2/24$ in (4.30) will result in EMPs violating them.

Here it is necessary to point out that the original proof of the Kramers–Kronig relations though commonly adopted contains a logical hole. The mean polarization of an effective medium is not induced by the mean field. On the contrary, the mean field results itself from the mean polarization. The last one is simply a spread microscopic polarization that is caused by the local fields at the particle centers.

Therefore, the correct way to derive Kramers–Kronig relations is to apply the causality, first, on the microscopic level, to the polarizabilities of the particles. Kramers–Kronig relations for the macroscopic parameters of media are implications. For the p-dipole polarization one can write instead of (1.7):

$$\mathbf{P}(\mathbf{r}, t) = \int_{-\infty}^{t} \overline{\overline{K}}_1(t-t') \mathbf{E}^{\text{loc}}(\mathbf{r}, t') dt'. \tag{4.31}$$

However, the CMLL formulas allow us to replace $\mathbf{E}^{\text{loc}}$ by the mean field $\mathbf{E}$ and to express $\mathbf{P}(\mathbf{r}, t)$ through the integral of $\mathbf{E}(\mathbf{r}, t)$ with another kernel $\overline{\overline{K}}$ that can be easily found through $\overline{\overline{K}}_1$.

This is possible because the CMLL formulas contain neither retardation terms nor high multipoles, nor the parameters of the external field. They only relate $\mathbf{E}^{\text{loc}}$, $\mathbf{E}$ and $\mathbf{P}$ locally and at the same time. If we assume that the external field is present in the relation between the local and mean fields, and substitute formula (4.30) into Eq. (4.31) the term $\mathbf{E}^{\text{ext}}$ from (4.30) will not allow the permittivity to be an analytic function in the upper half-plane of complex frequencies. The effective continuity requires that the local field is unambiguously expressed via the mean field and polarization.

As we have already understood, EMPs of media with WSD either satisfy the Kramers–Kronig relations or are useless. Therefore, the correct generalization of the CMLL formula for isotropic arrays is obviously (4.22) and not (4.30). So, the term with the factor $-(kd)^2/24$ in (4.30) must be an error of our model approximation. Recall that we have already omitted the term with $(kd)^2/6$ in Eq. (4.6). Now, we understand that all these terms should cancel out with the omitted remainder of the series in (4.11). It was not proved strictly, but it must be so.

Notice that the generalized CMLL formulas (4.22) and (4.23) can be easily extent to anisotropic lattices (as it was done for the static case in [55, 66, 67]). For the anisotropic lattice factor ξ in (4.22) will be a tensor.

4.2 On Boundary Conditions beyond the Metamaterial Frequency Region

4.2.1 Preliminary Remarks

For media with noticeable b-parameter the Maxwell boundary conditions are not sufficient for solving the boundary problems and one needs to derive some extra boundary conditions for the mean-field derivatives. However, below we concentrate on media with negligibly small b-parameter—media whose WSD manifests as the artificial magnetism and/or bianisotropy. For such media Maxwell's boundary conditions are sufficient for solving the boundary problems. We have already claimed that these conditions are granted by the covariance of MEs. This covariance really ensures the consistency in boundary conditions for the involved multipoles and effective surface charges and currents. However, up to now we have not yet specified at which boundary we will impose Maxwell's boundary conditions. We have not yet shown the microscopic content of these conditions. This will be done below.

Our effective medium is split onto orthorhombic or cubic unit cells. There is nothing physically different in the orthorhombic case with different periods $d_x \ne d_y \ne d_z$. Therefore, for simplicity of writing let us assume that the array is isotropic and the unit cell is cubic. Next, the small deviations of the particles positions from the lattice nodes also bring nothing new—our MEs are covariant and we may nicely refer the multipole centers of all particles to the nodes as it is shown in Fig. 3.1. If so it must be clear that the top surface of our effective-medium slab is not a plane on which the centers of the top particles lie (as it was in the Ewald theory) but the plane raised with respect to these centers by $d/2$. And the bottom surface of the effective slab is a plane distanced from the centers of the bottom particles by $d/2$ down. This means that our effective medium slab includes the integer number N of unit cells across it. Usually, bulk composites prepared namely in this way—the thickness Δ of the matrix is equal Nd. If it is larger, the effective medium slab still has the thickness Nd, and is sandwiched between two dielectric layers of total thickness $\Delta - Nd$. In all practical cases we may assume that the effective medium slab comprises an integer number of unit cells across it.

At these effective boundaries the tangential component of E_t and H_t of the mean field vectors **E** and **H** should not experience jumps. The same refers to the continuity of the normal components D_n and B_n of mean field vectors **D** and **B** across the boundaries. Although these jumps are expected in the case of multipole resonances, in this chapter we still develop the quasi-static theory. Now, two questions need to be answered: How do our mean fields and polarizations behave at the effective boundary, and which N is sufficient so as to treat the array as a bulk medium sample?

As we have already noticed $N = 2$ is not enough for this—the extinction principle and the CMLL formulas do not hold, and therefore, the bulk homogenization model is inconsistent. From another side, we have seen on the example of an array of dielectric spheres that $N = 4$ is an adequate number for considering the array as a sample of a bulk material. To clarify it, the same numerical simulations were done for $N = 1\ldots 10$. Analyzing the dependence on the incidence angle we have seen that a qualitative error arises if $N = 1$ and $N = 2$. If $N = 3$ the theory of WSD keeps adequate at least in the frequency range $(qd) < 1$. So, for arrays of non-resonant dielectric spheres with periods $d = 2a$ and $d = 1.5a$ and permittivities $\varepsilon_s = 10$ and $\varepsilon_s = 30$ the minimal amount of monolayers sufficient for the bulk homogenization is equal $N_{\min} = 3$. For other particles and geometries the needed $N_{\min}$ may be different.

In the present section, for simplicity, we restrict our analysis by the array of spheres and analyze only the impact of the boundary to the effective permittivity. The generalization of the suggested approach to particles of arbitrary shape and multipoles is straightforward due to the covariance of our MEs.

Let us better understand why the homogenization is inconsistent for composite slabs with small N. When we derive the set of EMPs we have no right to restrict the position of the observation point, otherwise MEs will be valid not everywhere and cannot be practically combined with Maxwell's equations for the mean fields. We have resolved this problem above locating the observation point inside the array. An arbitrary point inside the array can be considered as a multipole center of the nearest particle, and the covariant form of our MEs allowed us not to worry on the Maxwell equations on the effective boundary.

However, the model is consistent only if the averaging volume V_{av} covers the sufficient domain with microscopic polarization—the domain whose volume is equal to the volume V_p of a particle. This is so if the observation point is sufficiently distant from the effective boundary—at least by a. For the observation points located above the top boundary V_{av} does not comprise the domains with microscopic polarization. In the spatial region the MEs are those of free space. If the observation point is near the effective boundary of the composite the volume V_{av} may comprise only a fraction of V_p.

However, if the observation point is located exactly on the top surface our averaging volume comprises only one half of V_p and $<\mathbf{P}>$ at this point is twice as smaller as $<\mathbf{P}>$ in the bulk. This is point B in Fig. 4.2. This peculiarity of the surface polarization has nothing to do with the CMLL formulas or extinction principle, it is an inconsistency of our model. If the composite slab is sufficiently thick this inconsistency does not matter. However, if N is small, it matters and the application of the bulk homogenization model results in serious errors (see [94]). Below the concept of the effective-medium boundary will be revised so that to remove the inconsistency. Simultaneously, our model will become applicable to ultimately thin composite layers [95].

4.2.2 Drude Transition Layers as Effective Boundaries

Assume for simplicity of writing that the matrix holding a cubic lattice of spherical particles is free space. Then our effective medium results in $\varepsilon = 1$ for the observation point C in Fig. 4.2 and all points located above it. For this point the averaging volume does not intersect with any particle and the microscopic polarization inside V_{av} is zero. For points distanced from the plane $z = d/2$ farther than $a/2$ the mean polarization is identically zero.

If the observation point is point A in Fig. 4.2 whose vertical coordinate is $z_1 = d/2 - a/2$, the averaging volume covers the particle completely. The polarization at the observation point is the same as for points located beneath. So, for points distant to the depth from the plane $z = d/2$ more than by $a/2$ there is no sharp spatial

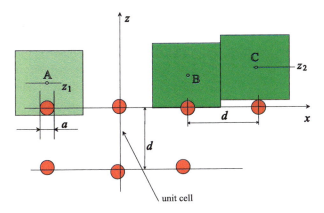

Figure 4.2 Averaging volumes for three different locations of the observation point (A, B and C) across the transition layer centered by the effective-medium boundary. The mean field varies slightly from point to point, but the mean polarization varies within this layer sharply.

variation in the mean polarization. If the mean field is uniform the polarization is uniform that corresponds to the standard concept of the effective permittivity ε.

However, for point B ($z = d/2$)the averaging volume covers one half of the top particle, and we have $\mathbf{P}(z = d/2) = \mathbf{P}(0)/2$. Over the interval $z_1 \equiv (d-a)/2 \leq z \leq (d+a)/2 \equiv z_2$, we may call the *transition layer*, the mean polarization varies from $\mathbf{P}(0)$ to 0, and this sharp variation can be approximated as a linear function:

$$\mathbf{P}(z) = \mathbf{P}(0)\frac{z_2 - z}{z_2 - z_1}, \quad z_1 < z < z_2. \tag{4.32}$$

Near the effective-medium surface $z = d/2$ we must consider the tangential polarization and the normal one separately and write:

$$\mathbf{D}_n(z) = \varepsilon_0 \varepsilon_n(z) = \varepsilon_0 \mathbf{E}_n(z) + \mathbf{P}_n(z), \tag{4.33}$$

$$\mathbf{D}_t = \varepsilon_0 \varepsilon_t(z) = \varepsilon_0 \mathbf{E}_t(z) + \mathbf{P}_t(z). \tag{4.34}$$

Here it is reasonable to stop and think what does it mean—the effective permittivity varying across the above-defined effective surface $z = d/2$.

In the theory by Ewald it was impossible. His effective permittivity referred to the spatially harmonic part of the lattice polarization.

However, Ewald's theory was a model of a lattice of point dipoles. Ewald did not take into account the finite size a of a particle, did not split the medium onto unit cells and his effective boundary was located at $z = 0$. His goal was to prove the extinction of the incident wave, i.e., to prove that the top layer of dipoles composed an effectively continuous interface.

The goal of our present study is different. Our homogenization model considers a finite array of finite-size particles and we do not need an extinction principle. We may introduce the effective boundary so that Maxwell's boundary conditions will be automatically satisfied for the mean field and auxiliary field vectors. When our observation point crosses this effective boundary our effective-medium model does not lose the validity, the same set of formulas simply describes the transition from the bulk medium to free space. So, we must have the z-dependent permittivity tensor, and our effective boundary is nothing but the Drude transition layer we have mentioned above several times.

The Ewald–Oseen peculiarity of the polarization of surface dipoles—their difference from the spatially harmonic wave polarization— it is simply not important for our permittivity tensor $\bar{\bar{\varepsilon}}(z)$. At the effective surface $z = d/2 = (z_1 + z_2)/2$ our effective permittivity varies very sharply because in accordance with (4.32) the polarization within the interval $[z_1, z_2]$ of thickness a drops from $\mathbf{P}(0)$ to zero. Meanwhile, within the interval $z_1 - \Delta < z < z_1$, where $\Delta = Nd$ is the effective thickness of the composite slab, the effective permittivity varies weakly. For all observation points located within this interval the averaging volume covers the volume $V_p = \pi a^3/6$. It can be that of a whole particle or two parts of two adjacent particles so that the total volume with microscopic polarization is V_p. The mean polarization across the effective medium slab is uniform if the mean field is uniform.

If $N = 2$ the polarizations of the top and bottom monolayers are (in the low-frequency limit) equivalent because both of them are interfacial layers of the effective-medium slab. It does not matter from which side the incident wave impinges the slab because the optical thickness of the slab is very small. If $N = 3$ the polarization of the central monolayer can be slightly different from that of the interfacial layers but the corresponding variation of $\mathbf{P}(z)$ across the bulk is negligibly small compared to the very sharp drop of $\mathbf{P}(z)$ across the very thin transition layer from $\mathbf{P}(0)$ to zero.

Even for optically thick effective layers, where the interfacial layers are polarized really differently from the polarization in the bulk, we may neglect the impact of the Ewald–Oseen peculiar polarization of surface dipoles compared to the sharp variation of our mean polarization across the transition layer. This variation is the main surface effect in our model. Therefore, it is reasonable to put $\varepsilon_t(z) = \text{const} \equiv \varepsilon$ within the interval $z_1 - \Delta < z < z_1$ describing the bulk of our array and write (4.34) in a form

$$\varepsilon_0 \varepsilon \mathbf{E}_t = \varepsilon_0 \mathbf{E}_t + \mathbf{P}_t \qquad z_1 - \Delta \leq z \leq z_1 \qquad (4.35)$$

for the bulk of our array, whereas for the transition layer we have

$$\varepsilon_0 \varepsilon_{\text{loc}}^t(z) \mathbf{E}_t = \varepsilon_0 \mathbf{E}_t + \mathbf{P}_t(z) \qquad z_1 \leq z \leq z_2. \qquad (4.36)$$

Here $\mathbf{P}_t(z)$ is described by (4.32).

Maxwell's boundary conditions require that $\mathbf{E}_t(z_1 + 0) = \mathbf{E}_t(z_1 - 0)$ if the effective surface is a plane. If the effective surface is the finite-thickness layer $z_1 < z < z_2$ the tangential mean field $\mathbf{E}_t$ is invariant across it. So, in Eq. (4.36) $\mathbf{E}_t$ is constant versus z. This uniformity of $\mathbf{E}_t$ allows us to obtain after substitution of (4.32) into (4.36) the coordinate dependence of the local permittivity:

$$\varepsilon_{\text{loc}}^t(z) = 1 + (\varepsilon - 1) \frac{z_2 - z}{z_2 - z_1}. \qquad (4.37)$$

Here, we assume for simplicity of writing that $\varepsilon_m = 1$. If the matrix occupies the interval $d/2 - \Delta < z < d/2$ we will have

$$\varepsilon_{\text{loc}}^t(z) = \varepsilon_m + (\varepsilon - \varepsilon_m) \frac{z_2 - z}{z_2 - z_1}.$$

For the vertical polarization the same approach is applicable. In this case the normal component of $\mathbf{D}$ must be uniform within the transition layer $z_1 < z < z_2$. Therefore, we have from (4.33):

$$E_n(z) = \frac{\varepsilon}{\varepsilon_{\text{loc}}^n(z)} E_n(z_1). \qquad (4.38)$$

Here we assume that the bulk permittivity ε is scalar because we consider the isotropic array. If the structure is anisotropic ε in (4.38) is different from ε in (4.37).

Substituting (4.32) and relation $P_n(z_1) = \varepsilon_0(\varepsilon - 1)E_n(z_1)$ following (2.6) into (4.38) we come to the following equation (when the host medium is free space)

$$\varepsilon_{loc}^n(z) = 1 + \frac{P_n(z)}{\varepsilon_0 E_n(z)} = 1 + \varepsilon_{loc}^n(z)\left(1 - \frac{1}{\varepsilon}\right)\frac{z_2 - z}{z_2 - z_1}. \quad (4.39)$$

The solution of this linear algebraic equation is simple:

$$\varepsilon_{loc}^n(z) = \frac{1}{1 - \left(1 - \frac{1}{\varepsilon}\right)\frac{z_2 - z}{z_2 - z_1}}, \quad z_1 < z < z_2. \quad (4.40)$$

So, even if the particles are isotropic (spheres) and form an isotropic structure such as a simple cubic lattice, the transition layer is anisotropic. Both tangential and normal components of the local permittivity tensor are equal unity on top of the transition layer and ε on its bottom, however, their coordinate dependence is essentially different. This anisotropy was noticed by Drude. Also, Drude has pointed out that the variation of the local permittivity holds in a optically small interval and needs the averaging. However, as we have already discussed the mistake of Drude was the postulate that the thickness of this interval is equal d—size of the averaging volume. And he averaged twice over the same interval—first, in order to calculate the local permittivity and, second, in order to average it.

We perform the first averaging over the volume $V = d^3$ whose size is d and the second averaging over the interval $[z_1, z_2]$ whose value is a. This averaging should be performed in accordance with the general rules for the tangential and normal components of the heterogeneous permittivity tensor [118]:[a]

$$\varepsilon_D^t = \frac{1}{a}\int_{z_1}^{z_2} \varepsilon_{loc}^t(z)\,dz \quad (4.41)$$

$$\varepsilon_D^n = a\left(\int_{z_1}^{z_2} \frac{dz}{\varepsilon_{loc}^n(z)}\right)^{-1}. \quad (4.42)$$

Here index D implies that this tensor permittivity is that of the Drude layer. The fact that a is the correct Drude layer

---
[a]The simplest interpretation of the Drude averaging rule for the normal component of the permittivity is a continuum of capacitances connected in series.

thickness, drastically changes his predictions for the reflection and transmission coefficients of a layer of a natural crystal or liquid medium. A layer filled with an anisotropic effective medium placed on two sides of a liquid medium sample noticeably changed the result in Drude's predictions for the Brewster angle [118] because this layer in the original theory had the thickness d. For natural liquids and molecular crystals $a \ll d$—molecules are much smaller than the inter-molecular distance. And if we replace in these predictions $d \to a$ the impact of the transition layers practically disappears. This fits the experimental data of [119]. In the next work dedicated to the check of the Drude formulas [120] Chandrasekhara Raman has ingeniously assumed that the predictions by Drude for the reflectance by a liquid are incorrect namely because Drude overestimates the thickness of the transition layer. Raman correctly assumed that this thickness is equal to a. The present study confirms this guess.

We have discussed the homogenization of the interface for the special case when the SD in the effective medium is absent. Similar speculations can be repeated in presence of the WSD and formulas similar to (4.37) and (4.40) derived for other EMPs. However, this is no need to do it if the medium is really bulk. Calculations for dielectric spheres have shown that the Drude transition effect is not important for sufficiently thick samples of composite media. Similarly, it is not important for arrays with WSD if $N \gg 1$. We have already achieved our goal—saw how Maxwell's boundary conditions are satisfied at the effective interface. The finite thickness of this interface and the variation of the EMP across it does not matter except an exotic case when the composite slab is ultimately thin. In the other cases we may nicely use the initial concept of the interface as a plane located at the height $d/2$ over the centers of the top particles.

4.2.3 Drude Transition Layer for Ultimately Thin Composite Slabs

In the case of the dielectric spheres the impact of the difference of ε_D^t and ε_D^n from ε is noticeable if $N =\leq 3$. Since such slabs are optically thin we may apply formulas (4.41) and (4.42) for the effective uniform permittivity of the whole slab replacing a by the slab physical thickness $\Delta = Nd$ enlarged by two transition layers of thickness a:

$$\varepsilon_{tot}^t = \frac{1}{\Delta} \int_{-(\Delta+a)/2}^{(\Delta+a)/2} \varepsilon_t(z)\,dz \qquad (4.43)$$

$$\varepsilon_{tot}^n = \Delta \left(\int_{-(\Delta+a)/2}^{(\Delta+a)/2} \frac{dz}{\varepsilon_n(z)} \right)^{-1}. \qquad (4.44)$$

Here $\varepsilon_t(z)$ and $\varepsilon_n(z)$ are uniform within $-\Delta/2 + a/2 < z < \Delta/2 - a/2$ and vary in accordance with (4.37) and (4.40) within the top Drude layer $\Delta/2 + a/2 < z < \Delta/2 - a/2$ and the bottom one $-\Delta/2 - a/2 < z < -\Delta/2 + a/2$.

This integration is doable analytically via tabulated integrals, and though the derivations are involved, the result is compact. The dependence on the size a of a molecule/particle disappears in this result:

$$\varepsilon_{tot}^t = \frac{2N}{2N+1}\varepsilon + \frac{1}{2N+1}, \qquad (4.45)$$

$$\varepsilon_{tot}^n = \frac{1}{1 - \left(1 - \frac{1}{\varepsilon}\right)\frac{\varepsilon_{tot}^t}{\varepsilon}}. \qquad (4.46)$$

Here ε is the bulk permittivity of a composite of spheres. Formulas (4.45) and (4.46) are correct if the host material is free space.

For the general case when arbitrary particles forming a uniaxial medium are located in a matrix of permittivity ε_m having the same thickness Δ as that of the effective medium, i.e., when the host medium slab comprises N unit cells across it, we have

$$\varepsilon_{tot}^t = \frac{2N}{2N+1}\varepsilon^t + \frac{\varepsilon_m}{2N+1}, \qquad (4.47)$$

and

$$\varepsilon_{tot}^n = \varepsilon_m \left[1 - \left(1 - \frac{\varepsilon_m}{\varepsilon^n}\right)\frac{\varepsilon_{tot}^t}{\varepsilon^t}\right]^{-1}. \qquad (4.48)$$

Similar formulas can be derived for the permeability of the particles possess a m-dipole moment. The same logics is applicable to the array of particles with a multipole response. However, corresponding formulas will be, probably, very cumbersome.

Numerical simulations for arrays of dielectric and metal spheroids have shown: if $N > 3$ both tangential and normal permittivities calculated using Eqs. (4.47) and (4.48) are close to their bulk values ε^t and ε^n and the Drude layers have a negligible impact for the power transmittance t and reflectance r of an effective slab. However, for $N \leq 3$ the impact of the Drude layers becomes noticeable and is really significant if $N = 1$.

In Fig. 4.3 circles and crosses depict the reflectance r calculated using the Fresnel-Airy formula for two cases of a simple cubic lattice of dielectric spheres: $N = 1$ (right panel) and $N = 3$ (left panel). Circles correspond to our model taking into account the Drude layers. We can see that this model perfectly agrees to rigorous numerical simulations (solid lines). Crosses correspond to the standard model in which the slab is filled by an isotropic medium, whose scalar permittivity is found by the Maxwell Garnett algorithm. The thickness Δ of the effective slab in this calculation is assumed to be equal to Nd. For the case $N = 1$ it is equal to the grid period d and is by 50% larger than the physical thickness a of the monolayer. It is possible to fit r for such an effective layer to the full-wave simulations considering the effective thickness as a fitting parameter. However, the agreement with the full-wave simulations for r is achieved when the effective thickness exceeds d and is accompanied by the disagreement for the power transmittance t. The bulk model ignoring the Drude layers either underestimates either the reflection or the transmission (though correctly predicts the Brewster angle).

As it is seen in Fig. 4.3, our model taking the Drude effect into account is successful. Our model does not require any fitting—all involved parameters are determined theoretically. In papers [95, 96, 98] one can find more examples confirming that the transition effect allowed us to increase the accuracy for both r and t coefficients for different ultimately thin composite layers.

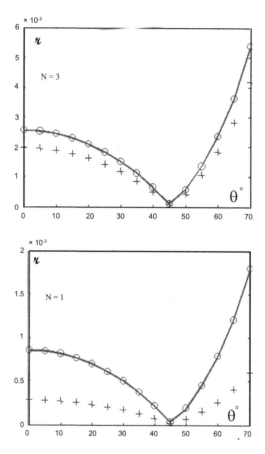

Figure 4.3 The angular dependence of the reflectance by a 3-layered (a) and 1-layered (b) lattice of spheres with permittivity $\varepsilon_s = 30$ located in free space and forming a cubic lattice with $d = 1.5a$. The normalized frequency is $kd = 0.1$. Crosses correspond to the simplistic model of the medium which ignores the effect of Drude transition layers. Circles correspond to our formulas (4.47) and (4.48). Solid lines represent the result of the full-wave calculation.

Of course, the description of a monolayer of particles, as a sample of a bulk medium is not physically perfect. It is possible to alternatively describe such a composite slab as an effectively planar structure. An optically dense grid of particles can be homogenized as a sheet of the surface polarization. However, the bulk description is simple and preferred by the majority of researchers. Therefore, it is reasonable to keep it when possible. We have seen above that it is possible if the constituents are not resonant. At frequencies where the particles are resonant, the present model fails and does not stand the comparison with full-wave simulations. Then it must be replaced by a model of a *metasurface* [20].

Chapter 5

Homogenization of Metamaterials with Artificial Magnetism

5.1 Dynamic Averaging

5.1.1 General Approach to Dynamic Averaging

Optically dense composites of resonant particles are metamaterials. In MMs the unit cell is obviously small compared to the wavelength in free space and in the host medium, but may be not very small compared to the effective-medium wavelength λ_{eff}. The wave in the effective medium may be several times shorter than in the host material due to the resonant shortening. As it was already mentioned, the concept of MM can keep working even if this size is as large as $\lambda_{\text{eff}}/3$. For a medium with optically substantial unit cells we cannot *a priori* ignore the polaritons induced by an incident wave at the MM interface. Polaritons may significantly change the microscopic field and Maxwell's boundary conditions for mean fields can be violated.

Of course, for microscopic fields Maxwell's boundary conditions hold obviously. However, the microscopic field comprises the contribution of polaritons, whereas mean field does not. Therefore,

---

Composite Media with Weak Spatial Dispersion
Constantin Simovski
Copyright © 2018 Pan Stanford Publishing Pte. Ltd.
ISBN 978-981-4774-83-3 (Hardcover), 978-1-351-16624-9 (eBook)
www.panstanford.com

the satisfaction of Maxwell's boundary conditions for microscopic fields in presence of polaritons contradicts to their satisfaction for mean fields. The medium may be still effectively continuous in the bulk but its interface is not that of a continuous medium. Analyzing the electromagnetic behavior of a MM slab, for a certain sheer of propagation angles we may observe no any feature of the electromagnetic discontinuity. Above we have already discussed these features and the violation of the Maxwell's boundary conditions for mean fields is not among them. This violation is the problem of our model. Let us therefore modify our model so that it describes such MMs as effectively continuous media in spite of the violation of Maxwell's boundary conditions for mean fields at their surfaces.

This new model of averaging does not needs to be applicable for any point of an array as we required above. There is no need to worry about the covariance of our MEs. It would be even unfair for such a MM because at the interface the effective medium is discontinuous. Therefore, let us restrict our homogenization model by an infinite array and consider the interface problem separately. This approach makes the new theory quite similar to the works by Ewald and Oseen, though our analysis of the interface problem is heuristic, i.e., much simpler.

In this approach, it is not obvious to analyze the difference between the local and the mean electric fields. Since the resonance of m-dipoles making them higher than the p-dipoles destroys the hierarchy of multipoles, Maxwell Garnett mixing rules may be not applicable. This is another argument in favor a completely new homogenization model.

Perhaps, some analogues of the CMLL formula still exist also for MMs but we do not need them. We will see that the homogenization of an infinite array is possible without mixing rules.

Now let us briefly discuss our heuristic approach to the boundary problem. The approach is based on the introduction of new transition layers. They are not Drude's transition layers, their purpose is to save the model of a continuous medium in the case when the Maxwell's boundary conditions are not satisfied for mean fields. Thus, new transition layers are needed for whatever thickness of the sample, even for a semi-infinite lattice. Jumps of the tangential components $\mathbf{E}_t$ and $\mathbf{H}_t$ of the mean electromagnetic field occur at the interfaces of the MM slab but these jumps can be spread across fictitious optically thin layers whose parameters are strongly contrast to both bulk MM and free space. In this way we preserve the model of continuous media and our original MM slab becomes a three-layer structure. A bulk MM slab is the central layer and two thin transition layers are at the interfaces. EMPs of both effective media—those of the bulk and those of the transition layers—can be found from the exact solution of the boundary problem.

Our theory of the infinite MM is a first-principle model. We start from the assumption that our constitutive particles can be described as p-dipoles and m-dipoles. In this model p-dipoles are excited by the local electric field (uniform within the particle) and m-dipoles are excited by the local magnetic field. In other words, we do not consider the BA arrays and restrict all effects of WSD by the most important one—artificial magnetism. Even this restricted study allows us to answer several fundamental questions concerning metamaterials as such.

Now, let us write a few words about the notations used below. This chapter is based on works [10–17] published later that [1]. Notations in these papers are different from those used in [1] and below we use namely the notations of these original works. Below a is the lattice maximal period (above it was d), g is the particle maximal size (above it was a). Brackets will be used in order to denote the averaging over the volumes different from that of the unit cell volume V.

5.1.2 New Definitions of the Mean Fields and Polarizations

To take into account the effects of SD is to take into account the retardation effects on the scale of a particle and/or a unit cell. Adopting the quasi-static averaging rule (2.12) we made our theory internally controversial. This controversy is insignificant for the non-resonant case when the corresponding error is small, but for resonant composites the rule (2.12) is not adequate and may result in significant errors.

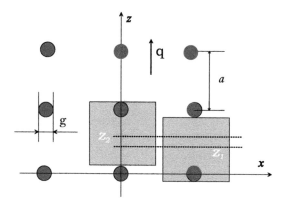

Figure 5.1 Averaging volumes for two different coordinates of the observation point. Two different crystal planes separated by the gap $a - g$ correspond to two averaging volumes centered at points z_1 and $z_2 = z_1 + g$, respectively. The simple averaging for defining the mean fields may be inapplicable in two cases: if the contribution of the internal fields into mean field is dominant (resonance regime) and if (qa) exceeds unity (photonic crystal regime).

Consider an infinite cubic lattice whose unit cells are shown in Fig. 5.1. Let an eigenwave with wave vector **q** travel along the z axis and let the particle maximal size g be of the same order as that of the period a. The requirement of optical density implies that $(qa) \ll \pi$, which can be specified as $(qd) < 1$ (as we have discussed in the previous chapter). Admit that the mean electric field is introduced by a quasi-static formula

$$\mathbf{E}(z) = \frac{1}{V} \int_V \mathbf{E}^{\text{true}}(\mathbf{r}')dV, \qquad (5.1)$$

where $V = a^3$ is the unit cell volume, and $\mathbf{r}'$ is the radius-vector of the integration point around to the observation point. If the definition (5.1) is consistent, this mean field is a plane wave with phasor $\mathbf{A}\exp(-jqz)$. Our goal is to show that (5.1) is not consistent for the resonant case.

Denoting the particle volume as V_p (for a sphere $V_p = \pi g^3/6$), the electric field inside the particle as $\mathbf{E}_{in}$ and the field in the host medium as $\mathbf{E}_{out}$ we may write

$$\mathbf{E}(z) = <\mathbf{E}_{in}> \frac{V_p}{V} + <\mathbf{E}_{out}> \left(1 - \frac{V_p}{V}\right). \qquad (5.2)$$

Here $<\mathbf{E}_{in}>$ is internal true field averaged over the particle volume V_p and $<\mathbf{E}_{out}>$ is the external true field averaged over the outer volume $V - V_p$. For the first observation point shown in Fig. 5.1 $z_1 = (a-g)/2$ and for the second point $z_2 = (a+g)/2$.

Two observation points z_1 and z_2 are separated by the gap a and the values of the eigenmode mean field taken at points z_1 and z_2 should differ by the phase shift (qg). If the particles are not resonant, their internal field (for metallic particles the role of $\mathbf{E}_{in}$ is played by the electric field at the particle surface) is of the same order of magnitude as the field outside them or even smaller. Together with the obvious condition $V_p \ll V$ (recall that we do not consider densely packed arrays!) this means that the second term in (5.2) dominates and the difference $\mathbf{E}(z_1) - \mathbf{E}(z_2)$ is approximately equal to $<\mathbf{E}_{out}>(z_1) - <\mathbf{E}_{out}>(z_2)$. Microscopic oscillations are practically absent outside the particles and $\mathbf{E}(z_1) - \mathbf{E}(z_2)$ is the value of the order $A[1 - \exp(-jqa)] \approx A(qg)$ as it must be. There is no inconsistency in this case.

Now, assume that still $(qa) < 1$ but particles are resonant. At the resonance the field is locally concentrated within the particle and the internal field of the particle strongly dominates over the external field. For complex-shape metal particles, such as SRRs, OPs and CPs, V_p is the volume of the particle envelope. In other words, the areas where the internal field is concentrated at the particle resonance (small gaps between metal patches, ends of the wires, etc.) are included into V_p. For dielectric particles at the electric Mie resonance the electric internal field is also much larger than the external field. For plasmonic nanoparticles the electric field is locally enhanced in some areas near the surface called *hot spots*. These areas are also included into V_p. In other words, for whatever resonant particles in Eq. (5.2) the first term dominates.

Neglecting $<\mathbf{E}_{out}>$ in (5.2) we obtain:

$$\mathbf{E}(z_1) - \mathbf{E}(z_2) \approx [<\mathbf{E}_{in}>(0) - <\mathbf{E}_{in}>(a)]\frac{V_p}{V}$$

$$= \frac{V_p <\mathbf{E}_{in}(0)>}{V}(1 - e^{-jqa}). \quad (5.3)$$

The left-hand side of (5.3) should be equal

$$Ae^{-jqz_1} - Ae^{-jqz_2} = Ae^{-jqz_1}(1 - e^{-jqg}),$$

whereas the right-hand side is equal
$$\frac{V_p <\mathbf{E}_{in}(0)>}{V}(1-e^{-jqa}).$$
In fact, the eigenmode phase shift between two adjacent particles centered at $z = d$ and at $z = 0$ is equal (qa). Therefore, the internal fields of these particles also differ by the phase factor $\exp(-jqa)$. Taking into account that $1 - \exp(-jqa) \approx jqa$ and $1 - \exp(-jqg) \approx jqg$ from (5.3) we obtain

$$<\mathbf{E}_{in}>(0) = \frac{gV}{aV_p}\mathbf{A}e^{-jqz_1}. \quad (5.4)$$

However, according to (5.2) we have

$$\mathbf{E}(0) \equiv \mathbf{A} \approx <\mathbf{E}_{in}>(0)\frac{V_p}{V}$$

that clearly contradicts to (5.3). This is so because the definition (5.1) is not consistent with the physical meaning of the mean field that must be a plane wave $\mathbf{A}\exp(-jqz)$.

For dielectric particles experiencing the magnetic Mie resonance the internal electric field averaged around the particle center does not dominate. Moreover, at the resonance frequency it vanishes. However, the quasi-static averaging rule is also not applicable in this case, since the same trouble with the consistent averaging keeps for the magnetic field.

Of course, the inconsistency in the mean field concept resulting from the averaging rule (5.1) is insignificant in the quasi-static limit when $(qa) \ll 1$. Then the phase shift $q(a - g)/2$ in (5.4) is negligibly small and the quasi-static mixing rule can be applicable even for a resonant lattice. In fact, the Maxwell Garnett model keeps applicable for optically dense arrays of very small plasmonic (Ag or Au) nanoparticles filling the metal glasses. However, if the diameter of a plasmonic nanosphere is substantial—larger than $g = 50$–60 nm—the averaging rule (5.1) turns out to be inconsistent. For such plasmonic lattices the quasi-static model will not work.

Briefly, for resonant lattices operating (though optically dense) beyond the quasi-static limit we have to modify our averaging procedure. To do it properly let us recall that we aim to homogenize only infinite regular lattices. Let us, therefore, introduce a weight function dependent on $\mathbf{q}$ into our averaging procedure and apply the theory of infinite lattices. Although the averaging procedure depends on **q** it may result in EMPs independent on **q**. Below we will see that it is really so for an important group of resonant lattices.

The true field of a wave propagating in an infinite lattice along z with the wave number q can be presented as the so-called *Bloch's series* of spatial harmonics [35–37]:

$$\mathbf{E}^{\text{true}}(x, y, z) = \sum_{mnp} A_{mnp} e^{-jqz + \frac{2\pi m}{a_x} + \frac{2\pi n}{a_y} + \frac{2\pi p}{a_z}},$$

where we assume that the lattice is orthorhombic but not obviously cubic (a_x, a_y and a_z can be different). This decomposition can be rewritten as

$$\mathbf{E}^{\text{true}}(\mathbf{R}) = A_{000} e^{-jqz} \left(1 + \sum_{mnp}' B_{mnp} e^{-j\mathbf{G}_{mnp} \cdot \mathbf{R}} \right), \quad (5.5)$$

where prime means the absence of the term $m = n = p = 0$ in the series, **G** is the vector with components $(2\pi m/a_x, 2\pi n/a_y, 2\pi p/a_z)$ and $B_{mnp} \equiv A_{mnp}/A_{000}$. If we multiply the right-hand side of (5.5) by $\exp(+jqz)$ all terms in the series will be periodic functions of coordinates. Therefore, integration over the unit cell volume V centered at the observation point **R** nullifies all of them but the zero-order term, and we obtain the mean field defined by such an integration in the needed form: $\mathbf{E} = A_{000} \exp(-jqz)$. This is our new automatically consistent definition of the mean field. In the general case when the wave with wave vector **q** propagates in the arbitrary direction with respect to the lattice axes we define the mean field as

$$\mathbf{E}(\mathbf{r}) = \frac{e^{-j\mathbf{q} \cdot \mathbf{r}}}{V} \int_V \mathbf{E}^{\text{true}}(\mathbf{r}') e^{+j\mathbf{q} \cdot \mathbf{r}'} \, dV, \quad (5.6)$$

Similarly, we will define the averaging of the true magnetic field. Since we neglect the quadrupoles and other multipoles, there is no problem anymore which vector is primary—**B** or **H**. For simplicity of writing, below we will average the microscopic vector $\mathbf{H}^{\text{true}}$ considering the macroscopic vector $\mathbf{B} = \mu_0 \mu \mathbf{H}$ as an auxiliary vector.

5.1.3 Averaging of Maxwell's Equations

Consider an infinite orthorhombic lattice of p–m dipole particles and write the Maxwell microscopic equations for it, which relate the

$\mathbf{E}^{true}$ and $\mathbf{H}^{true}$ strengths with the microscopic electric and magnetic polarizations $\mathbf{P}^{true}$ and $\mathbf{M}^{true}$ as follows:

$$\nabla \times \mathbf{E}^{true} = -j\omega(\mu_0 \mathbf{H}^{true} + \mathbf{M}^{true}),$$
$$\nabla \times \mathbf{H}^{true} = j\omega(\varepsilon_0 \varepsilon_m \mathbf{E}^{true} + \mathbf{P}^{true}). \quad (5.7)$$

Let a wave with wave vector $\mathbf{q}$ propagate along the z axis. Denoting the periods of the lattice by a_x, a_y, and a_z, respectively, we write the Bloch expansion of this wave as follows:

$$\begin{cases} \mathbf{E}^{true}(\mathbf{r}) \\ \mathbf{H}^{true}(\mathbf{r}) \end{cases} = \begin{cases} \mathbf{E}_0 \\ \mathbf{H}_0 \end{cases} e^{iqz} + \sum_{\mathbf{n} \neq 0} \begin{cases} \mathbf{E}_\mathbf{n} \\ \mathbf{H}_\mathbf{n} \end{cases} e^{i(qz + \mathbf{G}_\mathbf{n} \cdot \mathbf{r})}, \quad (5.8)$$

where the vectors $\mathbf{G}_\mathbf{n} \equiv (G_x, G_y, G_z) = (2\pi n_x/a_x, 2\pi n_y/a_y, 2\pi n_z/a_z)$ are the integer multiple vectors of the reciprocal lattice and $\mathbf{n}$ is the three-dimensional summation index, with the fundamental (zeroth) Bloch harmonic being single out from the total sum: $\mathbf{n} = (n_x, n_y, n_z)$, $n_{x,y,z} = \pm 1, \pm 2, \ldots$. The microscopic polarizations (electric and magnetic) can be expressed in terms of dipole moments $\mathbf{p}_0$ and $\mathbf{m}_0$ of an arbitrary chosen particle whose center is associated with the coordinate origin. In this homogenization model we may assume for simplicity of derivations that p- and m-dipoles are point ones. If we take into account the finite volume V_p occupying a part of the unit cell volume V, the result will not change. This is so because our MEs are applied below only for crystal planes $z = z_n$. In this case the averaging volume does not intersect our particles and their finite sizes do not matter.

Thus, in the present derivations we can use the model of point dipoles and write the microscopic densities of the dipole moments via three-dimensional delta-functions:

$$\begin{cases} \mathcal{P}(\mathbf{r}) \\ \mathcal{M}(\mathbf{r}) \end{cases} = \begin{cases} \mathbf{p}_0 \\ \mathbf{m}_0 \end{cases} e^{iqz} \sum_\mathbf{n} \delta(\mathbf{r} - \mathbf{r}_\mathbf{n}) \quad (5.9)$$

Convolute Eq. (5.7) with a certain kernel $K(r)$ defined in the entire infinite space. Then, from (5.7) two macroscopic Maxwell's equations follow, which are mathematically equivalent to them:

$$\nabla \times \mathbf{E} = i\omega(\mu_0 \mathbf{H} + \mathbf{M}),, \quad (5.10)$$
$$\nabla \times \mathbf{H} = -i\omega(\varepsilon_0 \varepsilon_m \mathbf{E} + \mathbf{P}), \quad (5.11)$$

where the used notations

$$\begin{cases} \mathbf{E}(\mathbf{r}) \\ \mathbf{H}(\mathbf{r}) \end{cases} = \int_{V_\infty} K(\mathbf{r} - \mathbf{r}') \begin{cases} \mathbf{E}^{true}(\mathbf{r}') \\ \mathbf{H}^{true}(\mathbf{r}') \end{cases} dV,$$

$$\begin{Bmatrix} \mathbf{P}(\mathbf{r}) \\ \mathbf{M}(\mathbf{r}) \end{Bmatrix} = \int_{V_\infty} K(\mathbf{r}-\mathbf{r}') \begin{Bmatrix} \mathbf{P}^{\text{true}}(\mathbf{r}') \\ \mathbf{M}^{\text{true}}(\mathbf{r}') \end{Bmatrix} dV$$

are our new definitions of the mean field and polarizations. It is clear that the quantities defined in this way are our new macroscopic fields and polarizations, respectively, if the kernel K is chosen in accordance with (5.6). Namely, choose the kernel as follows:

$$K(\mathbf{r}) = \begin{cases} (1/V)e^{i\mathbf{q}\mathbf{r}} = (1/V)e^{iqz}, & \mathbf{r} \in V \\ 0, & \mathbf{r} \notin V \end{cases}, \quad (5.12)$$

where $V = a_x a_y a_z$ as above, is the unit volume of the lattice. Substituting (5.9) and (5.12) into (5.10) and (5.11), and calculating the obtained elementary integrals, we can easily see that our mean fields are nothing other than zeroth Bloch harmonics of the true field:

$$\begin{Bmatrix} \mathbf{E}(\mathbf{r}) \\ \mathbf{H}(\mathbf{r}) \end{Bmatrix} = \begin{Bmatrix} \mathbf{E}_0(\mathbf{r}) \\ \mathbf{H}_0(\mathbf{r}) \end{Bmatrix} e^{iqz}. \quad (5.13)$$

For macroscopic polarizations, averaging yields piecewise constant functions of coordinates rather than harmonic functions. However, it is not important for us, since we do not take care on the covariance of our MEs.

Substituting (5.9) and (5.12) into (5.10) and (5.11), we obtain also the following result for the polarizations:

$$\begin{Bmatrix} \mathbf{P}(\mathbf{r}) \\ \mathbf{M}(\mathbf{r}) \end{Bmatrix} = \begin{Bmatrix} \mathbf{p}_0/V \\ \mathbf{m}_0/V \end{Bmatrix} e^{iqz_n}, \quad z_n - \frac{a_z}{2} < z < z_n + \frac{a_z}{2}. \quad (5.14)$$

Here $z_n = a_z n_z$ are z-coordinates of the lattice nodes. With substitutions (5.13) and (5.14), the Maxwell macroscopic equations (5.10) and (5.11) referred to the crystal planes $z = z_n$ take the following form:

$$qH(z_n) = \omega\varepsilon_0\varepsilon_m E(z_n) + \omega P(z_n), \quad (5.15)$$

$$qE(z_n) = \omega\mu_0 H(z_n) + \omega M(z_n). \quad (5.16)$$

Here we take into account that the mean fields have linear polarizations and in the present case (propagation along a crystal axis) are mutually orthogonal, e.g., $\mathbf{E}$ is x-polarized and $\mathbf{H}$ is y-polarized.

Now let us introduce the EMPs of our effective magneto-dielectric material taking into account that the microscopic p- and m-dipoles $\mathbf{p}_0$ and $\mathbf{m}_0$ are proportional to the electric and magnetic local fields, respectively. Our MEs are as simple as

$$\mathbf{D} = \varepsilon_0 \mathbf{E} + \mathbf{P}, \quad \mathbf{B} = \mu_0 \mathbf{H} + \mathbf{M}.$$

Then the definitions of EMPs ε and μ through mean fields and polarizations become traditional ones:

$$\mathbf{P} = \varepsilon_0 (\varepsilon - \varepsilon_m) \mathbf{E}, \tag{5.17}$$

$$\mathbf{M} = \mu_0 (\mu - 1) \mathbf{H}. \tag{5.18}$$

Substituting (5.17) and (5.18) into (5.15) and (5.16) we easily obtain:

$$q^2 = k^2 \varepsilon \mu, \quad Z_w \equiv \frac{E(z_n)}{\eta_0 H(z_n)} = \sqrt{\frac{\mu}{\varepsilon}}, \tag{5.19}$$

where $\eta_0 = \sqrt{\mu_0/\varepsilon_0}$ is the wave impedance of free space.

Expressions (5.19) uniquely relate the EMPs with the propagation constant q of the wave and its impedance Z_w. Earlier, they were known for continuous media and for p-m lattices were obtained only in the framework of the quasi-static model. Here, the same relations were derived without any restrictions on the frequency range. The frequency range where Eqs. (5.19) can be useful is that where EMPs introduced by formulas (5.17) and (5.18) are relevant.

The question of their relevance is not trivial. In fact, the wave number $\mathbf{q}$ and wave impedance Z_w are not characteristics of the effective medium. They are properties of the lattice eigenmode! We have to find this eigenmode and after we have found $\mathbf{q}$ and Z_w, we will determine our permittivity and permeability. If everything is done correctly and the effective medium is really continuous, these EMPs defined by Eqs. (5.17) and (5.18) really describe the electric and magnetic responses of the medium. If these parameters are, for example, antiresonant, our model fails. Reasons of the possible model failure have been already mentioned and will be discussed below in details.

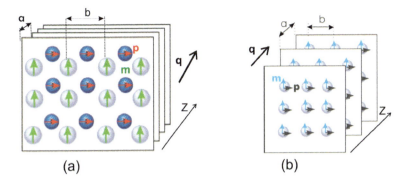

Figure 5.2 An infinite orthorhombic lattice of p-m particles is represented as a set of crystal planes. Two cases—separate p-dipoles and m-dipoles (a) and p-m-dipoles unified in one particle (b)—correspond to the same relations between the wave parameters and EMPs.

5.1.4 Generalization of the Model

Notice that all derivations presented above can be easily generalized to the case when p- and m-dipoles of the lattice are spatially separated as it is shown in Fig. 5.2a. Of course, in this case the wave number q and the wave impedance Z_w will be different from those calculated for the case depicted in Fig. 5.2b, but relations (5.19) keep valid. This follows from the fact that the functions $\mathbf{P}(z)$ and $\mathbf{M}(z)$ are constant within the given cell.

Next, the presented averaging model can be extended to the case where the wave propagates obliquely to the coordinate axes of the lattice. Then the factor $\exp(-jqz)$ should be replaced by $\exp(-j\mathbf{q} \cdot \mathbf{r})$. The averaging procedure of Maxwell equations with this kernel results in a rather simple equation for the wave vector, which is better to express via the normalized wave vector $\mathbf{n}$, defined as $\mathbf{q} = k_0 \mathbf{n}$. Here $k_0 = \omega\sqrt{\varepsilon_0 \mu_0}$ is the wave number of free space. Skipping very long derivations let us write the final result in the index form:

$$n^2 I_{ij} - n_i n_j = \varepsilon_{ik} \mu_{kj}. \qquad (5.20)$$

This is the well-known characteristic equation of an arbitrary anisotropic magneto-dielectric medium [53], whose proof had been earlier known only for continuous media. Here, it was obtained within the framework of a dynamic model for a lattice of particles.

Therefore, the locality of bulk EMPs in the right hand side of (5.20) is not clear a priori.

As to the wave impedance, it turns out to be a tensor, and simple closed-form equations are obtained for it only in the case when the eigenwave **q** is either TE or TM with respect to a certain crystal plane, e.g., plane $(x - y)$ and the grids of p-m particles located in the crystal planes parallel to $(x - y)$ have the same period $d_x = d_y = b$. In both cases when the p-m dipoles are spatially separated as shown in Fig. 5.2a and when they exist in each particle as shown in Fig. 5.2b such a lattice is a dynamic analogue of the uniaxial continuous medium with optical axis z (if both periods $a \equiv d_z$ and b are sufficiently small compared to the wavelength $\lambda_{\text{eff}} = 2\pi/q$). The lattice is then characterized by the uniaxial dyadic effective permittivity $\varepsilon = \varepsilon_t(\mathbf{x}_0\mathbf{x}_0 + \mathbf{y}_0\mathbf{y}_0) + \varepsilon_n\mathbf{z}_0\mathbf{z}_0$ and similar permeability $\mu = \mu_t(\mathbf{x}_0\mathbf{x}_0 + \mathbf{y}_0\mathbf{y}_0) + \mu_n\mathbf{z}_0\mathbf{z}_0$. The same result was obtained for the wave impedance in both cases shown in Fig. 5.2a,b:

$$Z_w^{TM} = \sqrt{\frac{\mu_t}{\varepsilon_t}}\sqrt{1 - \frac{q_t^2}{k_0^2 \varepsilon_n \mu_t}} \qquad (5.21)$$

and

$$Z_w^{TE} = \sqrt{\frac{\mu_t}{\varepsilon_t}}\bigg/\sqrt{1 - \frac{q_t^2}{k_0^2 \varepsilon_t \mu_n}}, \qquad (5.22)$$

where q_t is the component of the wave vector tangential to the plane $(x - y)$. The wave impedance in (5.21) and in (5.22) is defined as follows:

$$Z_w \equiv \frac{E_t(z_n)}{\eta_0 H_t(z_n)}, \qquad z_n = na, \quad n = 0, \pm 1, \pm 2, \ldots.$$

Here the index t denotes the tangential (to the crystal plane) component of the mean field. Eqs. (5.21) and (5.22) as well as their special case, Eqs. (5.19), reproduce the well-known equations for the wave impedance of a continuous uniaxial magneto-dielectric in the formalism where the wave impedance is referred to the transversal plane (that orthogonal to the optical axis) [55].

The system of Eqs. (5.20) in the uniaxial case yields to two simple relations:

$$q_n^2 \equiv q^2 - q_t^2 = k_0^2 \varepsilon_t \mu_t, \quad q_t^2 = k_0^2 \varepsilon_n \mu_n. \quad (5.23)$$

Fixing the certain propagation direction with respect to the lattice coordinate at a given frequency we may solve the lattice eigenmode problem for both TE and TM eigenwaves—find four scalar parameters q_t, q_n, Z_w^{TM} and Z_w^{TE}. Then from Eqs. (5.21), (5.22), and (5.23) we express four scalar EMPs to be evaluated—ε_t, ε_n, μ_t and μ_n. This problem is solvable even for the case when p- and m-dipoles are spatially separated not only in the crystal planes parallel to $(x - y)$ but also along z.

However, this solution is very involved and below we will explore only the case when the eigenwave propagates along one of the coordinate axes, e.g., along z. If the composite is uniaxial and z is its optical axis, both Z_w^{TM} and Z_w^{TE} yield to Z_w and we may retrieve from (5.19) only ε_t and μ_t. These two components of the dyadic EMPs of a uniaxial composite are introduced above by (5.17) and (5.18). For the purposes of the present book the analysis of these transversal components of the resonant effective permittivity and permeability is sufficient.

To conclude this subsection let us discuss which explicit MMs can correspond to the sketch with separate p- and m-dipoles shown in Fig. 5.2a. A resonant magnetic moment is impossible without a p-dipole response that can vanish only at isolated frequencies. However, the p-dipole can be negligibly small compared to the m-dipole in the resonance band of latter. One of two constitutive particles in the two-phase composite depicted in Fig. 5.2a can be a plasmonic nanosphere (in the optical version of the MM) or a short inductively loaded wire dipole (in its microwave realization). Another one can be a resonant SRR or a dielectric sphere, experiencing the magnetic Mie resonance (both of them can be implemented in both microwave and optical versions). Also, the two phases (p-dipole and m-dipole ones) can be implemented by two different dielectric spheres—one experiencing the electric Mie resonance and another experiencing the magnetic Mie resonance in the frequency range of our interest. We have already mentioned this MM suggested in [152] and pointed out that at the resonance of m-dipoles of the small spheres the quasi-static p-dipoles of the big spheres can be taken into account through the modified permittivity of the host medium. So, there are several realizations of the p-m lattices with the separate p-dipoles and m-dipoles.

All formulas derived above keep valid for a 2D lattice when the constitutive particles are aligned cylinders of arbitrary cross section. In this case we have to restrict our consideration by the propagation of the waves in the plane orthogonal to the cylinders. For the dielectric cylinders the resonances correspond to the TM-polarization of the eigenmode. The electric field of the TM mode and its magnetic field excite in the cylinders a transversal p-dipole and longitudinal m-dipole per unit length, respectively. In the case of the metallic cylinders the resonances arise when the cylinders have a complex shape of the cross section, e.g., their cross section is shaped as a split ring or as a Swiss roll. These resonances are also excited by the TM waves. If the electric field is polarized along the axis of a solid metal or dielectric optically thin cylinder, there are no resonances.

Here, it is difficult to abstain of mentioning the so-called *wire media*—MMs of thin metal cylinders whose most interesting properties are related namely with the electric polarization along the wires. This is one of two known types of MM whose unusual properties have nothing to do with the resonance of a constitutive inclusion. Wire media are referred to MMs due to a collective effect—strong SD that does not vanish even at low frequencies—the quasi-static limit does not exist for wire media at all.[a] Wire medium is a MM for which the non-local effective permittivity not only makes sense—it is very useful (like the non-local permittivity of semiconductor crystals). Additional boundary conditions for wire medium layers were established in the closed analytical form. Therefore, this permittivity helps not only to study the eigenmodes but also to solve boundary problems [191].

---

[a]The same property is inherent to a class of non-resonant MM called *hyperbolic metamaterials*. These MMs are practically important at optical frequencies and are implemented as stacked media of dielectric and plasmonic nanolayers [70, 71]. Some authors refer to wire media as a special type of hyperbolic MMs.

5.2 Effective Material Parameters of the Resonant p-m-Lattices

This section, based on papers [11, 13–17] is dedicated to the search of EMPs of orthorhombic p-m lattices through the parameters of the eigenmode. The electromagnetic interaction in such lattices was studied analytically for both cases illustrated by Fig. 5.2a,b where the crystal plane comprises both p- and m-dipoles and for the case when the grids of p-dipoles alternate with those of m-dipoles. Rather accurate solutions were found analytically for all these cases and the accuracy of these solutions was studied numerically by comparison with full-wave simulations. This theory is rather mathematically difficult and involved. Here we do not stay on its details and concentrate on the results relevant for our purposes. Recall that our purposes are homogenization of the lattice in terms of local EMPs.

5.2.1 Wave Number of a p-m-Lattice

In this subsection we consider two geometries of an orthorhombic p-m lattice: a set of alternating p- and m-grids as shown in Fig. 5.3a and a set of equivalent grids where p- and m-dipoles are combined as shown in Fig. 5.3b. Above, we assumed that every crystal plane comprises both p- and m-dipoles and distinguished two cases— when they refer to the same particle or when they refer to different particles of the same grid. Here we unite the structures depicted in Figs. 5.2a,b into one case illustrated by Fig. 5.3b because the dispersion equation is the same. The structure of alternating grids depicted in Fig. 5.3a is characterized by a different dispersion equation.

In this subsection we will see how to find the wave number q of an eigenmode propagating along z for two types of a p-m lattice illustrated by Fig. 5.3a,b. Local electric and magnetic fields in an infinite lattice in absence of the external excitation is the interaction field produced by the p-dipoles and m-dipoles which are all similar to those induced in the reference particle (or two reference particles in the cases shown in Figs. 5.2a and 5.3a).

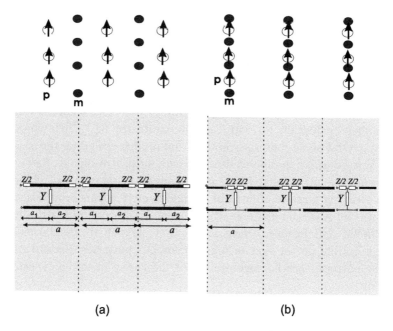

Figure 5.3 Crystal planes of an infinite orthorhombic lattice of p-m particles can be alternating grids of p-dipoles and m-dipoles (a). They can be also identical grids in which p-dipoles and m-dipoles are both present (b). These two types of a p-m lattice are equivalent to corresponding types of a periodically loaded transmission line shown on the bottom panel.

The difference between p- and m-moments induced in different grids orthogonal to z is only in the phase imposed by the propagating eigenmode wave vector $\mathbf{q} = q\mathbf{z}_0$. The phase shift between two adjacent grids of the same type is qa. Denoting the electric and magnetic polarizabilities of p-dipoles and m-dipoles, respectively as a_{ee} and a_{mm} we can write following relations for the p-moment $\mathbf{p} = p\mathbf{x}_0$ and m-moment $\mathbf{m} = m\mathbf{y}_0$ induced in the reference particle (or in two reference particles if they are different):

$$p\mathbf{x}_0 = a_{ee} E^{\text{loc}} \mathbf{x}_0, \quad E^{\text{loc}} = B_e p + B_{em} m, \qquad (5.24)$$
$$m\mathbf{y}_0 = a_{mm} H^{\text{loc}} \mathbf{y}_0, \quad H^{\text{loc}} = B_m m + B_{me} p. \qquad (5.25)$$

Parameters B_e, B_m, B_{em} and B_{me} are *interaction factors* of the lattice, which depend on the lattice geometry, on the frequency ω and on the wave number q. The four interaction factors can be all found after we have expressed all p-dipoles and m-dipoles of the lattice through the reference dipole moments—p-dipole p and m-dipole m using the Bloch quasi-periodicity conditions:

$$\mathbf{p}(N) = \mathbf{p}e^{-jNqa}, \qquad (5.26)$$

$$\mathbf{m}(N) = \mathbf{m}e^{-jNqa}. \qquad (5.27)$$

In the case when the p-grid and m-grid are alternating as in Fig. 5.3a, the number N of a p-grid and that of a m-grid are counted from two reference grids, respectively. If the p-grid and the m-grid are combined in the same crystal plane N is the number of one crystal plane. In both cases $N = 0$ correspond to a reference grid.

Using (5.26) and (5.27) one can analytically find four interaction factors of the p-m lattice as functions of the lattice geometry, ω and q. In fact, only two of them are independent—B_e is proportional to B_m and B_{em} is simply equal to B_{me}. These two independent interaction factors are expressed through the dynamic dipole sums for which the special mathematical methods were developed in cited papers and offer a strict solution in the form of a series. These series sums were estimated analytically and calculated numerically. For optically dense orthorhombic lattices these two series are rapidly converging and with high accuracy can be approximated by a reasonable number of terms.

Further, it is evident that a system of two source-free equations following from (5.24) and (5.25), namely

$$p(1 - a_{ee}B_e) - ma_{ee}B_{em} = 0, \qquad (5.28)$$

$$m(1 - a_{mm}B_m) - pa_{mm}B_{me} = 0, \qquad (5.29)$$

has the nonzero solution if and only if its determinant is zero. This condition results in the general dispersion equation:

$$[1 - a_{ee}(\omega)B_e(\omega, q)][1 - a_{mm}(\omega)B_m(\omega, q)]$$

$$- a_{ee}(\omega)a_{mm}(\omega)B_{em}(\omega, q)B_{me}(\omega, q) = 0. \qquad (5.30)$$

Expressions for interaction factors (in the form of series) are different for two geometries shown in Fig. 5.3a,b. These expressions are also different for two implementations of the combined p-m grid illustrated by Fig. 5.2a,b. However, the dispersion equation is unique for two last cases, only the effective parameters entering it are different.

In the most important when the lattice period a along the wave propagation direction is nearly equal or larger than the internal period of the crystal plane b the near-field interaction between two adjacent crystal planes is much weaker than the near-field interaction of the dipole scatterers in the same grid. Then the general, rather involved, expressions for the interaction factors simplify and can be written in the closed form. In this case the general dispersion equation (5.30) for both cases depicted in Fig. 5.3a,b reduces to a form that coincides with the dispersion equation of a periodically loaded transmission line.

Namely, for the structure illustrated by Fig. 5.3b Eq. (5.30) takes form:

$$\cos qa = \cos k_m a \left(1 + \frac{YZ}{4}\right) + \frac{j}{2}\left(Y + Z + \frac{YZ^2}{4}\right) \sin k_m a, \quad (5.31)$$

where $k_m = k_0\sqrt{\varepsilon_m}$ is the wave number of the matrix.

Parameters Y and Z entering this dispersion equation have a clear physical meaning, illustrated by Fig. 5.3. Namely, Y is a grid admittance of an individual grid of p-dipoles normalized to the admittance of the host material $Y_m = \sqrt{\varepsilon_0\varepsilon_m/\mu_0}$). Notice that a crystal composed by only p-dipole grids was studied in [190], where the dispersion equation

$$\cos qa = \cos k_m a + \frac{jY}{2} \sin k_m a,$$

was derived. It is easy to see that this equation is the special case $Z = 0$ of our Eq. (5.31).

The dimensionless grid admittance Y in (5.31) is a coefficient relating the tangential component of the transversely averaged electric field $\mathbf{E}_t^{TA}$ taken at the grid plane with the averaged surface current $\mathbf{J}_s$ induced in it: $\mathbf{J}_s = YY_m\mathbf{E}_t^{TA}$. Parameter Y is independent on q and can be found from the solution of an auxiliary problem— the plane wave incidence on an *individual* grid. To find Y one has to use the following definition of the transversal averaging:

$$\mathbf{E}^{TA}(\mathbf{r}) = \frac{e^{-j\mathbf{k}_t \cdot \mathbf{r}}}{S} \int_S \mathbf{E}^{\text{true}}(\mathbf{r}')e^{+j\mathbf{k}_t \cdot \mathbf{r}'} \, dS, \quad (5.32)$$

where $\mathbf{k}_t$ is the tangential component of the incident wave vector and S is the grid unit cell area. In the present case when the eigenmode propagates along z the auxiliary problem corresponds to the normal incidence. Then $k_t = 0$, and integration in (5.32) is a simple averaging over the unit cell surface. In this case Y is a scalar value.

Notice that the grid admittance YY_m in the classical book [80] is called the sheet admittance and defined via the jump of the tangential magnetic field across the grid: $[\mathbf{H}_t^{TA}(z = +0) - \mathbf{H}_t^{TA}(z = -0)] = YY_m \mathbf{n} \times \mathbf{E}_t^{TA}$, where $\mathbf{n}$ is the unit normal to the grid plane $z = 0$. The equivalence of these two definitions follows from Maxwell's boundary condition $\mathbf{J}_s = \mathbf{n} \times [\mathbf{H}_t(z = +0) - \mathbf{H}_t(z = -0)]$.

Parameter Z is dual to Y. It can be called the dimensionless magnetic grid impedance of an individual crystal plane of m-dipoles. It is dimensionless since normalized to the host medium impedance $Z_m = 1/Y_m = \sqrt{\mu_0/\varepsilon_0 \varepsilon_m}$. Namely, Z is introduced via the effective magnetic surface current $\mathbf{J}_s^{(m)}$ or, equivalently, via the jump of the tangential component of the transversely averaged electric field across the grid: $\mathbf{n} \times [\mathbf{E}_t^{TA}(z = +0) - \mathbf{E}_t^{TA}(z = -0)] \equiv \mathbf{J}_s^{(m)} = ZZ_m\mathbf{H}_t^{TA}(z = 0)$. It is evident that this jump if $\mathbf{H}_t$ and this magnetic surface current result from tangentially polarized magnetic dipoles in the grid.

If these m-dipoles are tangentially oriented, as it is shown in Fig. 5.3a, and there is no tangentially oriented p-dipoles, $\mathbf{H}_t^{TA}$ is continuous across the grid. The grid is characterized by the Z parameter and its Y parameters is zero. This is the case dual to the case of the p-dipole lattice studied in [190]. The dispersion equation for this case is the same with the replacement $Y \to Z$.

If the grid combines both p- and m-dipoles, as shown in Fig. 5.3a, $\mathbf{H}_t^{TA}(z = 0)$ in the definition of Z is an arithmetic mean value of two averaged values. The first averaged value $\mathbf{H}_t^{TA}(z = +0)$ is obtained by the averaging (5.32) of the true magnetic field at one side of the grid. The second one $\mathbf{H}_t^{TA}(z = -0)$ is the transversely averaged magnetic field at another side of the grid. Then $\mathbf{H}_t^{TA}(z = 0) = [\mathbf{H}_t^{TA}(z = +0) + \mathbf{H}_t^{TA}(z = -0)]/2$. Similarly, in the definition of Y we have $\mathbf{E}_t^{TA}(z = 0) = [\mathbf{E}_t^{TA}(z = +0) + \mathbf{E}_t^{TA}(z = -0)]/2$.

Parameters Y and Z can be expressed via the electric a_{ee} and magnetic a_{mm} polarizabilities of individual particles forming the grid [16]:

$$(jY)^{-1} \approx \frac{b^2 \varepsilon_0 \varepsilon_m}{k_m} \left(\frac{1}{a_{ee}} - \frac{jk_m^3}{6\pi \varepsilon_0 \varepsilon_m} \right)$$
$$- \frac{1}{4} \left(\frac{\cos 0.6954 k_m b}{0.6954 k_m b} - \sin 0.6954 k_m b \right), \quad (5.33)$$

$$(jZ)^{-1} \approx \frac{b^2 \mu_0}{k_m} \left(\frac{1}{a_{mm}} - \frac{jk_m^3}{6\pi \mu_0} \right)$$
$$- \frac{1}{4} \left(\frac{\cos 0.6954 k_m b}{0.6954 k_m b} - \sin 0.6954 k_m b \right). \quad (5.34)$$

These simple relations are valid only in the special case when $a \geq b$. Only in this case the series responsible in the interaction factors for the near-field interaction between the adjacent grids can be neglected.

Also, it worth to notice that the response of a grid to the incident wave splits onto two separate parameters—Y and Z—because the array of the in-phase p-dipoles does not interact with the array of m-dipoles located in the same plane. In Fig. 5.2 all p-dipoles are stretched along x and m-dipoles are along y, i.e., all are tangential to the grid plane. Such p-dipoles do not create the tangential magnetic field in the grid plane and such m-dipoles do not create the tangential electric field. Therefore, they cannot interact. However, even if the constitutive particle have complex shape and their p- and m-dipoles possess the normal component, they also cannot interact due to the problem symmetry. All z-directed p-dipoles of the grid in the case of the normal incidence are equivalent and in the grid plane $z = 0$ their magnetic fields cancel out. So, for the propagation along the crystal axis one always can describe the response of a crystal plane by a pair of parameters Y and Z which do not depend on q.

For the oblique propagation of the eigenmode, it also possible to split the response of a grid to two parameters Y and Z if we separately consider the propagation of TE-polarized waves and TM-polarized waves. However, in this case Y and Z become depending on the tangential component of **q** and the model becomes difficult. Therefore, in this book we consider only the normal propagation of the wave in a regular metamaterial—the eigenmode propagates along one of its crystal axis.

Then to find Y and Z entering Eq. (5.31) it is enough to solve the problem for an individual grid impinged by a normally incident plane wave. For grids with in-plane anisotropy Y and Z are strictly speaking tensor values even for the normal incidence. However, the eigenwaves with x- and y-polarizations of $\mathbf{E}$ are not coupled in the case under study. Therefore, for a given polarization of the wave Y and Z are scalar values. Moreover, in this chapter we restrict our model by the case of in-plane isotropy.[b]

If the p-dipole grid and m-dipole grid are alternating as shown in Fig. 5.3a, the dispersion equation is different. In the simplest special case when $a_1 = a_2 = a/2$ the dispersion equation is as follows:

$$\cos qa = \cos k_m a + j\left(\frac{Y+Z}{2}\right)\sin k_m a - \frac{YZ}{2}\cos^2\frac{k_m a}{2} \quad (5.35)$$

If $a_1 \neq a_2$ we have to solve a more involved equation

$$\cos qa = \cos ka + j\sin k_m a \left(\frac{Y}{4} + \frac{Z}{4}\right)$$

$$-\sqrt{\frac{YZF}{4} - \sin^2 k_m a \left(\frac{Y}{4} - \frac{Z}{4}\right)^2}. \quad (5.36)$$

Here a new notation is introduced:

$$F = 2\cos qa \cos k_m a_1 \cos k_m a_2 - \cos^2 k_m a_1 - \cos^2 k_m a_2.$$

In the theory of periodically loaded transmission line both Eqs. (5.31) and (5.35) are known. These equations correspond to shunting loads of admittance Y and series loads of impedance Z. The equivalent lines are shown in the bottom panel of Fig. 5.3a. In this figure the unit cell is introduced splitting every grid of m-dipoles onto two halves. This allows us to present a unit cell of the lattice by a symmetric T-circuit. One may redefine the unit cell shifting it by $a/2$. This will modify the equation (5.35) because the unit cell becomes non-symmetric. However, the solution q of the modified equation cannot depend on the choice of the unit cell and definitely keeps the same.

---

[b]The generalization to the case when the grid is in-plane anisotropic is straightforward: Y^{xx} and Z^{yy} are given by modified formulas (5.33) and (5.34), where $a_{ee} = a_{ee}^{xx}$, $a_{mm} = a_{mm}^{yy}$, and $b = \sqrt{d_x d_y}$. Similarly, the pair (Y^{yy}, Z^{xx}) is found.

Again, in our dispersion equations Y and Z are parameters of an individual p-grid or an individual m-grid. This model is applicable—formulas (5.33) and (5.34) are correct—if the near-field interaction between the adjacent grids is negligibly small compared to the near-field interaction in the grid. In other words, the grids interact with one another by plane waves.

Numerical studies have shown that the normalized value of the z-period (ratio a/λ) of the lattice is not relevant for this assumption. The near-field interaction between the adjacent grids along z is either strong or negligible depending on the ratio a/b. Practically, if $(b/a) > 2$–3 the near-field interaction between the grids along z is strong. In this case the interaction factors of the lattice are more complicated and do not allow us to yield them to the consolidated response of a single grid. If $(b/a) \le 1$–2 the near-field interaction between the grids is negligible. This is so because the in-plane interaction is that of the in-phase dipoles. It is much stronger than the interaction between the adjacent grids whose dipoles have different phases.

In this subsection we have seen how to find q. Let us now find Z_w. Then we will be able to determine bulk EMPs of our lattices from Eqs. (5.19).

5.2.2 Wave Impedance of a p-m-Lattice

From (5.28) or (5.29) one may easily derive two equivalent expressions for the ratio of the p-dipole to the m-dipole moment:

$$\frac{p}{m} = \frac{1}{B_{me}}\left(\frac{1}{a_{mm}} - B_e\right), \quad \frac{m}{p} = \frac{1}{B_{em}}\left(\frac{1}{a_{ee}} - B_m\right). \quad (5.37)$$

From definitions of the bulk material parameters we have

$$p = VP = \varepsilon_0 V(\varepsilon - \varepsilon_m)E, \quad m = VM = \mu_0 V(\mu - 1)H,$$

that gives the relationship between the ratio p/m and wave impedance Z_w of the effective medium:

$$\frac{p}{m} = \frac{\varepsilon_0(\varepsilon - \varepsilon_m)}{\mu_0(\mu - 1)} Z_w. \quad (5.38)$$

Substituting interaction factors from the theory of p-m lattices into the first relation in the system (5.37) and equating it to the right-hand side of (5.38) we obtain for the case of equidistant alternating grids (see Fig. 5.3a):

$$Z_w = \eta \frac{\gamma k_0 + q}{\gamma q + k_0 \varepsilon_m}, \qquad (5.39)$$

where $\eta = \sqrt{\mu_0/\varepsilon_0}$ is the free-space impedance and the following notation is used

$$\gamma = \sin\frac{qa}{2}\left[\frac{j(\cos k_m a - \cos qa)}{Y \cos \frac{k_m a}{2}} - \sin\frac{k_m a}{2}\right]^{-1}. \qquad (5.40)$$

Z_w only seemingly does not depend on Z. It depends on Z via q. An alternative equation for the parameter $\gamma \equiv \eta p/m$ can be derived using the second relation in the system (5.37):

$$\gamma = \left[\frac{j(\cos k_m a - \cos qa)}{Z \cos \frac{k_m a}{2}} + \sin\frac{k_m a}{2}\right]\left(\sin\frac{qa}{2}\right)^{-1}. \qquad (5.41)$$

Equating (5.40) and (5.41) one obtains the dispersion equation (5.35).

Similar expressions for the coefficient γ were derived for the lattices of combined grids (containing both p- and m-dipoles). Reproduction of all these formulas is not the target of the present book. Notice only that formula (5.39) keeps valid for combined grids.

5.2.3 Effective Material Parameters of an Infinite Lattice

Now we have everything we need in order to calculate the EMPs of an infinite lattice of resonant p- and m-dipoles. The key question is are these parameters local or depending on q. As it has been already discussed, if their frequency behavior does not agree with the requirements of the locality (causality and passivity), e.g., is antiresonant, these EMPs are useless. The failure of the homogenization model means that either the medium is not effectively continuous or the homogenization model developed above is not appropriate for it. If the homogenization model gives physically sound EMPs, the medium is effectively continuous and our model is appropriate for it because our ε by definition describes the electric response of a continuous medium and our μ describes its magnetic response.

As to the correct prediction of the reflective and refractive properties of our MM, this question should be postponed until we have considered the transition layers.

It is useful to introduce the refraction index $n \equiv q/k_0$ of the lattice and to rewrite formulas (5.19) with the substitution of (5.39) in the form

$$\varepsilon = \frac{n(n\gamma + \varepsilon_m)}{\gamma + n}, \quad \mu = \frac{n(\gamma + n)}{n\gamma + \varepsilon_m}. \quad (5.42)$$

Applying these relations for calculating the EMPs we have to be attentive. The propagation factor q representing the numerical solution of the lattice dispersion equation is attributed to the Brillouin zone $0 < q < \pi/a$. From the theory of photonic crystals [35, 37] it is known that in the second passband the true propagation factor K—that of the dominating spatial harmonic in the plane-wave expansion (5.8) of the lattice eigenmode—is equal to $K = 2\pi/a - q$. For the third passband it is equal $K = 2\pi/a + q$. For the fourth passband $K = 4\pi/a - q$, etc. Due to the resonances of the constituents these passbands can occur at low frequencies—below the Bragg reflective resonance is expected, i.e., below the frequency at which $a = \lambda/2$ (here λ is the wavelength in the matrix, which can be much larger than $\lambda_{\text{eff}} \equiv 2\pi/K$).

At the frequencies of these high-order passbands the lattice can be either an effectively continuous or an effectively discrete medium. The continuity can be seen from the analysis of the lattice dispersion and isofrequencies. If it is effectively discrete it is a photonic crystal, and we do not take care on it. If it is effectively continuous we have to identify the refractive index n with the dominating spatial harmonic in the expansion (5.8).

Next, we have to correctly calculate the sign of n. The eigenmode energy is by definition propagating forward. This means that an eigenmode excited by an external source obviously propagates from the source. Therefore, the correct formula for the refractive index must allow it to be negative. This is so in the case of a backward wave.

The theory of the lattice frequency dispersion implies that the wave number K of the dominating harmonic is positive. In fact, in the second passband $q < 2\pi/a$, in the fourth passband $q < 4\pi/a$, etc. To obtain the backward wave we have to take into account the direction of the group velocity. In the assumption that $K > 0$ the group velocity is calculated as $V_g = \partial \omega/\partial K$. So, we have find $K(\omega)$ through $q(\omega)$, then we have to find the group velocity $V_g = \partial \omega/\partial K$ and if it is negative it means that the phase and group velocities have opposite directions. So, calculating the refractive index n in the first passband we have to use the formula $n = \text{sign}(V_g)q/k_0 = \text{sign}(\partial \omega/\partial K)q/k_0$. For the second passband $n = \text{sign}(V_g)[(2\pi/k_0 a) - q]$, etc.

5.2.3.1 A two-phase lattice of dielectric spheres

In the beginning of 21st century, an idea of an isotropic doubly negative medium looked so exciting that the word *metamaterial* was invented especially in order to designate namely this hypothetic medium (see in [72, 73, 193]), generalizing the idea of the Veselago medium—that with $\varepsilon = \mu = -1$. In works [72] and [73] only an anisotropic variant of a doubly negative MM was suggested and studied. Later, many anisotropic doubly negative MMs operating in different frequency ranges were suggested, whereas the isotropic design turned out to be more challenging.

An idea of isotropic artificial magnetism in a dielectric composite of spheres with magnetic Mie resonance was suggested long ago—in work [152]. In a rather recent paper [154] this idea was extent in order to obtain an isotropic doubly negate material at microwaves. The suggested structure is shown in Fig. 5.4a. It is a two-phase composite of ceramic spheres with two different diameters. Bigger spheres have radius $R_e = 3.18$ mm, smaller spheres have radius $R_m = 2.28$ mm. The permittivity of the ceramic (both big and small spheres) is equal $\varepsilon_s = 44(1 - j10^{-4})$. Ceramics with such a complex permittivity in the range 5–20 GHz are available on the world market. The lattice period is $d_z = 2a = 20$ mm, where a is the distance between the centers of two adjacent (small and big) spheres. The host medium in this numerical example is free space $\varepsilon_h = \varepsilon_0$.

For the propagation along one of crystal axes (axis z) all grids comprise both p- and m-dipoles and are identical with respect to their Y and Z parameters. These parameters are not affected by the lateral shift a of alternating grids in the $(x - y)$ plane. Therefore, the model of identical crystal planes (p-m grids) keeps adequate for such a two-phase lattice. So, in our electrodynamic model the period of the lattice entering (5.31) is equal to $a = 10$ mm.

The electric and magnetic polarizabilities of the spheres were calculated using formulas (10–15) of [154]. These formulas are taken from the Mie theory, we have already discussed them. Spheres with radius $R_e = 3.18$ mm experience the individual electric Mie resonance at 10 GHz and those with $R_m = 2.28$ mm experience at this frequency the magnetic Mie resonance. As it was already noticed that the big ("electric") spheres though possess the nonzero m-dipole polarizability in the range 9–11 GHz, it is negligibly small compared to that of the small ("magnetic") spheres. We have also pointed out that the electric polarizability of "magnetic" spheres in this range can be taken into account in a form of the modified host medium permittivity ε'_m:

$$\varepsilon'_m = \varepsilon_m(1 - F_m) + \varepsilon_s F_m \approx 1.54,$$

where F_m is the volume fraction of the "magnetic" spheres. It can be checked that this simple formula fits the Maxwell Garnett mixing rule in the present case when $F_m \ll 1$ and $\varepsilon_s \gg \varepsilon_m$. With this modified matrix permittivity in mind we can consider our "magnetic" spheres in the range 9–11 GHz as m-dipoles with the magnetic polarizability a_{mm} and our "electric" spheres as p-dipoles with the electric polarizability a_{ee} (in the same host).

The dispersion in the lattice was calculated for the case of the normal propagation using the homogenization model expressed by Eq. (5.31) and by full-wave simulations using CST Studio software. The obtained lattice dispersion for this case is presented in Fig. 5.4b. The theoretical results are in agreement with full-wave simulations. Moreover, in this broadband plot both curves visually coincide and the agreement seems to be perfect.

Up to this point our study of the structure gives the same result as in [154]. However, the fine structure of the dispersion in the resonance band of the spheres is not visible in Fig. 5.4b. The resonant part of the dispersion curve is separately shown in Fig. 5.5a. Namely, both real and imaginary parts of the dimensionless wave number (qa) are shown versus frequency in the band 9.8–10.2 GHz. Although the full-wave simulations qualitatively match our model also in this much finer scale, the shifts of all characteristic

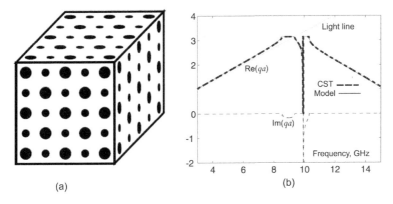

Figure 5.4 A two-phase cubic lattice of small ceramic ($\varepsilon_s \approx 44$) spheres from work [154] experiencing two basic Mie resonances near 10 GHz (a) and the broadband dispersion diagram of this lattice (b) obtained using our model and using the CST simulations.

frequencies make the visual analysis of the resonant dispersion difficult if we depict both analytical and numerical results in the same plot. In Fig. 5.5a we present only the analytical model of the resonant dispersion. It differs drastically from that obtained in [154] using the Maxwell Garnett model.

For better understanding of the wave properties we also show the frequency dispersion of the refractive index in Fig. 5.5b. In Fig. 5.5a we observe the second passband of the structure at frequencies 9.5...9.9 GHz. Here, the decrease of q versus frequency does not mean the backward wave because $K = \pi/a - q$ and the refractive index n in this band is positive and growing versus frequency. The hybrid resonance band of both p-dipole of the "electric" spheres and m-dipole of the "magnetic" ones occupied the interval 9.901...9.936 GHz and consists of two parts. In the high-frequency part—at 9.928...9.936 GHz—we observe the third passband of our lattice. Here the refractive index is negative and this passband corresponds to the backward wave. In work [154] it was claimed that in the corresponding band the effective permittivity and permeability are both negative and, consequently, the medium is an effectively continuous. Since it is a cubic lattice it is, definitely, an isotropic and low-loss doubly negative material [154].

In fact, in accordance with Fig. 5.5h in the third passband of the lattice the imaginary part of the refractive index is small, i.e., attenuation of the backward wave (at least in the range 9.928 . . . 9.931 GHz) is low. However, the dispersion of the refractive index in Fig. 5.5b is essentially non-Lorentzian unlike that calculated in [154]. Instead of the resonant losses around the frequencies 9.901–9.928 GHz at which the real part of n nullifies we observe the Bragg mode—the second stopband—at 9.890 . . . 9.928 GHz and the third passband at 9.928 . . . 9.936 GHz. Clearly, these are features of the strong spatial dispersion. It is not surprising. The lowest Bragg stopband occurs at lower frequencies—at 8.5 . . . 9.5 GHz. It is seen in Fig. 5.4b. So, the backward wave we observe has nothing to do with that in the doubly negative medium. It is a backward eigenmode of a photonic crystal. Such backward waves have been known since the 1950s where they were engineered for microwave generators called backward-wave tubes. Such backward waves do not promise us exciting properties inherent to backward waves in a continuous medium [193].

The fact that the backward-wave regime holds at frequencies above the first Bragg lattice resonance can be easily estimated without any accurate theory and full-wave numerical simulations. At low frequencies the array of small spheres can be homogenized by averaging its permittivity with that of the host. This gives for the effective permittivity of a medium in which the big spheres are located $\varepsilon_m = \varepsilon_s F_m + (1 - F_m) \approx 1.44$. This effective medium has the refractive index $n_m \approx 1.2$. The distance between the big spheres along the axis z is $2a = 20$ mm, and the first Bragg resonance of such a lattice corresponds to the effective wavelength $\lambda_{\text{eff}} = \lambda/n_m = 4a$. It gives the first Bragg wavelength $\lambda = 48$ mm, i.e., the lowest Bragg frequency should be located at 6.25 GHz. Of course, this is a very rough estimation but it clearly indicates that the Mie resonances of the MM and the corresponding backward wave regime occur in the frequency region where the medium cannot be effectively continuous.

Notice that the if we smooth the real part of the resonant refractive index n presented in Fig. 5.5b and broaden a little two peaks of its imaginary part the result will be very similar to the plot of n presented in Fig. 8 of [154]. In this work, the refractive

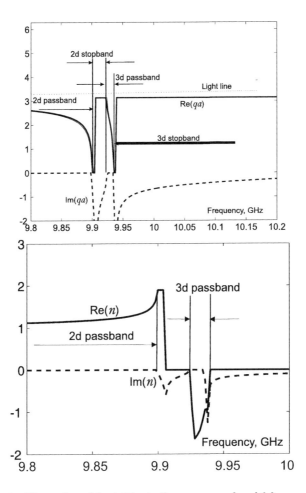

Figure 5.5 Dispersion of the lattice in the resonance band (a), an enlarged part of the plot Fig. 5.4b, and dispersion of the lattice refractive index (b).

index was numerically simulated using the Nicholson-Ross-Weir retrieval procedure for a finite-thickness slab. It turned out to be visually rather close to the broadband Lorentzian plot of n calculated using the Maxwell Garnett model. This similarity was, probably, the reason why the authors of [154] believed that their structure was a doubly negative MM. However, an attentive visual inspection allows one to detect in the retrieved refractive index of [154] the Bragg stopband and other features of the strong SD. The

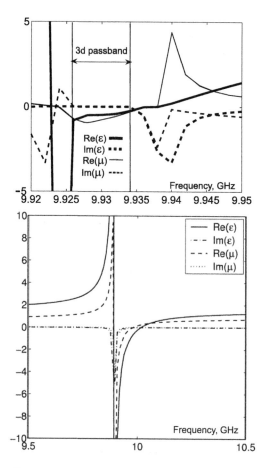

Figure 5.6 Effective permittivity and permeability of the lattice calculated in accordance with the present model (a) and using the Maxwell Garnett model for the two-phase composite (b) as it was done in work [154].

magnification of the corresponding plots completely destroy the impression of the agreement between the quasi-static model and full-wave simulations. This difference is especially evident from our Fig. 5.5b, where the frequency plot of n is not smooth. Regarding the bulk EMPs of this presumably doubly negative isotropic MM, the difference of the present model from the quasi-static model of [154] is even more drastic. The EMPs resulting from our dynamic modeling are depicted in Fig. 5.6a. It is seen that they have nothing to do with purely Lorentzian EMPs calculated in [154]. The last ones are reproduced in Fig. 5.6b.

Within the resonant stopbands which are specially marked in Fig. 5.6a, our EMPs have evidently no physical meaning. As to the third passband, that of backward wave, our model gives both negative ε and μ, which fits the usual understanding of the backward wave in the isotropic continuous medium. However, this doubly negative result is a trap for inexperienced researchers—in fact, our model gives meaningless permittivity and permeability also in this range. In fact, our permeability is not monotonously growing as it must do in the range of negligibly small magnetic losses. It has a local minimum there, and this frequency behavior is not compatible with Lorentzian dispersion. Our EMPs violate the Kramers–Kronig relations, and this violation points to the strong SD in the effective medium. There is nothing against our model in this result, but the fact that a homogenization model is not applicable to photonic crystals.

5.2.3.2 A lattice of omega dimers

A numerical example considered above is an example of misinterpreted numerical data where the array of p-m particles is only seemingly a doubly negative metamaterial. There are, definitely, effectively continuous arrays supporting backward waves such as the initial design solution [72] with wires and SRRs. Of course, this structure operates as a doubly negative MM only for the waves propagating orthogonally to the wires. Here we present another example of such a structure formed by particles of the same type. It has an advantage with respect to [72] in what concerns the polarization of the propagating wave. In [72, 73] the backward wave corresponded only to the waves whose electric field was polarized along the wires, the wave with the electric polarization across the wires and the magnetic polarization along them did not interact with the effective medium. Our structure is free of this drawback. Consider, a lattice composed by OPs arranged as it is presented in Fig. 5.7. The structure comprises two mutually orthogonal arrays of parallel Omega dimers. Every dimer comprises two resonant OPs performed of copper and located on two sides of a thin lossless

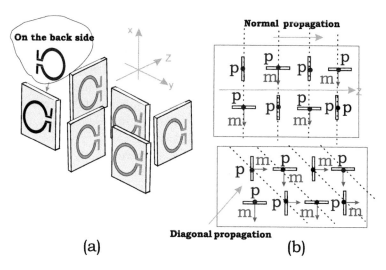

Figure 5.7 Uniaxial lattice of paired OPs. (a) Side view of the structure. Two Omega-particles are located on two sides of a thin dielectric plate so that the bianisotropy is cancelled. (b) Top view of the lattice for two cases: the normal propagation and the diagonal propagation. Dashed lines show the selected grid planes coinciding with the wave fronts. Bold points show the direction of electric dipole moments.

dielectric plate. Two OPs in every dimer are oriented so that the dimer's bianisotropy cancels out. Dimers are resonant p-m particles and the whole lattice represents a magneto-dielectric MM.

In the plane $(y - z)$ this lattice is isotropic. In order to prove it in [16] two cases of the propagation were considered: the normal propagation along the axis x or y and the diagonal propagation in between these axes. It is simple to modify the existing model so that to apply it for the case when the propagation is diagonal. The effective grid planes for the diagonal and normal cases of propagation are shown in Fig. 5.7b by dashed lines. We can see that for the diagonal propagation the lattice is still a set of mutually equivalent p-m grids orthogonal to the propagation direction. The in-plane period for this case increases ($b \to b\sqrt{2}$) and the main period a decreases ($a \to a/\sqrt{2}$). This modification still allows us to neglect the near-field interaction between two adjacent crystal planes.

Now let us discuss the issue which may potentially influence to the in-plane isotropy. For two cases of propagation the amount of m-dipoles in a crystal plane is different. For the normal propagation only one half of all Omega-dimers acquire the m-dipole moment. In the case of the diagonal propagation all dimers possess the m-dipole. At a first glance, this difference is important since the amount of magnetic moments duplicates. However, the rings of the OPs form the angle $\pi/4$ with the direction of the magnetic field that brings the factor $1/\sqrt{2}$ into a_{mm}. Moreover, the components of **m** normal to the grid planes do not contribute into the eigenmode. Only tangential components of **m** contribute that brings one more factor $1/\sqrt{2}$ into a_{mm}. Therefore, no significant difference arises for the magnetic response of a grid comparing two directions of propagation. Moreover, the change in the grid electric admittance and magnetic impedance caused by the enlarged internal period b of the grid, its impact in the interaction factors of the lattice is almost compensated by the impact of the reduction of a—the lattice period along the propagation path. As a result, two sets of EMPs calculated for two propagation directions turn out to be rather close.

In Fig. 5.8 the EMPs extracted from Z_w and n for the normal and the diagonal cases of the wave propagation, respectively, are depicted. Neglecting a weak additional resonance arising for the diagonal propagation at 4.3 GHz one can consider these EMPs as a unique set of parameters. At least, it so for the backward-wave frequency region 3.96–4.29 GHz. We have an in-plane isotropic effectively continuous doubly negative MM. This example shows that the strong SD in the backward-wave region is not an obvious feature of doubly negative MMs.

A similar comparative study of the normal and diagonal propagation was done in [16] for a two-phase lattice of dielectric spheres considered above. In that case, the result was as discouraging as the study above. The propagation in the direction diagonal with respect to a cubic domain shown in Fig. 5.4a resulted in a set of antiresonant EMPs which had nothing to do (except the absence of the physical meaning) with those obtained above for the propagation along the crystal axis. In the diagonal case the negative values of the effective permittivity and those of the effective permeability were also obtained, but they were obtained

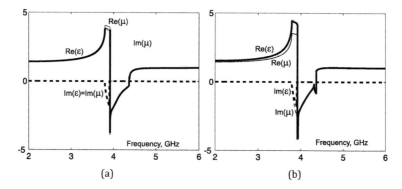

Figure 5.8 Effective material parameters extracted for the uniaxial lattice of Omega dimers (thick lines for the complex permittivity, thin lines for the complex permeability). (a) Normal propagation. (b) Diagonal propagation. Neglecting a weak resonance arising for the diagonal propagation and a small difference in $\mathrm{Re}\mu$ one can consider these EMPs as a unique set of parameters.

negative in different frequency bands. Therefore, no backward-wave frequency region was obtained at all. In the band 9.01–9.36 GHz where the backward wave propagates in the normal direction the propagation in the diagonal direction was prohibited by a directional stopband. For a seemingly isotropic lattice this huge anisotropy of the electromagnetic properties is more than sufficient indication of the strong SD. This study has shown once more: A two-phase lattice of dielectric spheres with $\varepsilon_s \approx 44$ is not an effectively continuous doubly negative material. In the range where the quasi-static model predicts the backward-wave regime is a self-resonant photonic crystal.

Of course, an effective medium of Omega-dimers depicted in Fig. 5.7a, though effectively continuous and isotropic for waves propagating in the $(x - y)$ plane, is not a fully isotropic doubly negative material. In order to transform it into an isotropic medium in work [195] one more sub-array of the Omega-dimers was added to this lattice. However, the 3D isotropy was not proven in that work. In paper [196] it was shown that this lattice was not isotropic and a better design solution, also based on OPs, was suggested that resulted in an isotropic doubly negative materials. To conclude, an

Omega-particle is a suitable building blocks of a doubly negative microwave MM (at least, in the theory, since the experimental verification of was [196] not done). The only drawback of such MMs are quite substantial optical losses in the backward-wave region. In Fig. 5.8 the imaginary parts of the effective permittivity and permeability in the backward wave region are not seen, but they are nonzero and correspond to the imaginary part of the complex negative refractive index $\text{Im}(n) = -(1\dots3) \times 10^{-2}$. Notice that in these calculations the dielectric was ideally lossless. For the 3D isotropic implementation of the Omega-lattice the resonant losses in copper were not calculated. Most probably, they will be higher than in the lattice of dimers studied above.

Isotropic doubly negative MMs are more important in the optical range. Such MMs were also suggested and theoretically studied in several works. Judging upon full-wave numerical simulations one may prefer the design solution suggested in work [197]. It is a simple cubic lattice of mutually touching core-shell nanoparticles with Ag cores and shells of intrinsic amorphous silicon. For such lattices with properly chosen geometric parameters the impact of the strong SD is minimized, and the isotropy turns out to be nearly perfect. However, this MM as any other resonant doubly negative composite manifests substantial losses in the backward-wave frequency band. The decay in the backward-wave band is higher by an order of magnitude than that in the lattice of Omega-dimers studied above. This drawback makes such a lattice not suitable for the imaging purposes.

This book is not concentrated on doubly negative MMs and we do not stay on the perfect imaging. However, it is worth to mention that in order to achieve the truly perfect imaging the MM must be ideally lossless. To obtain though imperfect but essentially subwavelength imaging in the far zone of an object using a layer of a doubly negative medium, one has to manage the issue of optical losses reducing them to very low values. This is hardly possible exploiting the Lorentzian resonance of a scatterer. Therefore, for subwavelength optical imaging other physical mechanisms have been developed (see [194]). As to doubly negative MMs, they keep interesting for researchers in view of other physical effects in them predicted in [3]. However, the practical impact of all these effects is lower than that of the far-zone subwavelength imaging. This is why the financial investments into this chapter of MMs keep insufficient for the experimental validation of the design solution suggested in [197].

5.2.4 Standard Retrieval of Bulk EMPs from the Scattering Matrix of a MM Slab

In the literature the term "dynamic homogenization of metamaterials" is often allegedly referred to as the heuristic electromagnetic characterization of a MM slab treated as a uniform bulk medium. We have already mentioned that this characterization is commonly performed using the Nicholson-Ross-Weir (NRW) method [198–201]. Here we discuss this method is more details. This is a following procedure. One assumes that a composite layer illuminated by a normally incident plane wave behaves as an effectively continuous magneto-dielectric medium. For the normal incidence the scattering matrix of the medium slab contains two scalar complex components—reflection R and transmission T coefficients. These coefficients refer to the phasor of the incident wave electric field taken at the front interface of the slab. Coefficients R and T can be measured or numerically simulated. For a regular layer composed by a rather small integer number N of monolayers they can be found easily. In this case the internal periodicity allows one to reduce the problem of an infinitely extended slab to the problem of N unit cells stretched along the propagation direction. This small array is confined at the sides by perfect magnetic and electric walls whereas the front and rear sides of the small array are corresponding slab interfaces.

If the lattice unit cell is symmetric, its T and R are the same for the illumination of an effective slab from both front and rear sides. These two complex numbers are uniquely related (by the commonly known Fresnel–Airy formulas) with two other complex numbers referring to the interior of the slab—its effective refractive index n and its characteristic impedance Z_c, usually normalized to the impedance of free space η. These two numbers are then found from R and T and attributed to the entire layer of thickness of $D = Na$. This is the first stage of the method.

The Fresnel–Airy formulas describing R and T or a continuous non-chiral medium, impinged by a plane wave normally incident from free space, are as follows:

$$R = \frac{R_\infty(1 - e^{2ik_0 nD})}{1 - R_\infty^2 e^{2ik_0 nD}}, \quad T = \frac{e^{ik_0 nD}(1 - R_\infty^2)}{1 - R_\infty^2 e^{2ik_0 nD}}, \qquad (5.43)$$

In formulas (5.43), coefficient R_∞ is interpreted as the reflection coefficient of the wave impinging the half-space of the same medium:

$$R_\infty = -\frac{1 - Z_c}{1 + Z_c}. \qquad (5.44)$$

In the second stage, one calculates the tangential components of the effective-medium tensors. In order to distinguish the retrieved EMPs from the true EMPs of the MM we will denote the retrieved ε and μ as ε_eff and μ_eff, respectively. The NRW algorithm was developed in order to characterize natural magneto-dielectric media at microwaves. Therefore, it comprises by default an assumption that the characteristic impedance Z_c and the refractive index n retrieved from R and T are the wave impedance Z_w of the effective medium and its refractive index, respectively. If it so, we have

$$n^2 = \varepsilon_\text{eff} \mu_\text{eff}, \quad Z_c = \sqrt{\frac{\mu_\text{eff}}{\varepsilon_\text{eff}}}. \qquad (5.45)$$

If this assumption is applicable to MMs we really have to treat ε_eff and μ_eff as the tangential components of the effective medium tensors.[c] However, it is questionable is this assumption really applicable to a realistic MM within the band of its magnetic resonance.

In works [8, 9] one may find unproven speculations that the relations (5.45) is an obvious implication of the effective continuity of MMs. It is really so, if and only if the MM is continuous together with its interface. However, in this book we have repeatedly enquired that a MM can be effectively continuous in its bulk whereas its interface is not continuous. The violation of Maxwell's boundary conditions makes (5.45) inapplicable.

---

[c] If the effective medium is isotropic, which is rarely the case, these two retrieved EMPs are scalars.

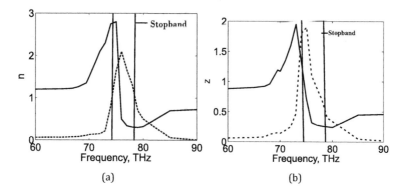

Figure 5.9 Frequency dependencies of the retrieved parameters of a lattice of infrared SRRs that reproduce the results of [8]: (a) refractive index n; (b) characteristic impedance Z_c. Here and below, the continuous curves correspond to the real parts of the retrieved parameters and the dashed curves—to the imaginary parts. The time dependence of the electromagnetic field in this example is $\exp(-i\omega t)$, different from that adopted in the rest of this book.

If we disregard this issue we may extract ε_{eff} and μ_{eff} inverting the Fresnel–Airy formulas:

$$q = \pm \frac{1}{D}\text{acos}\left(\frac{1-R^2+T^2}{2T}\right) + \frac{2\pi m}{D} = k_0\sqrt{\varepsilon_{\text{eff}}\mu_{\text{eff}}}, \quad (5.46)$$

$$Z_c = \pm\sqrt{\frac{(1+R)^2-T^2}{(1-R)^2-T^2}} = \sqrt{\frac{\mu_{\text{eff}}}{\varepsilon_{\text{eff}}}}. \quad (5.47)$$

The key point of the NRW algorithm is the correct choice of the sign and the unknown integer number m in formula (5.46). These choices are different for different frequency intervals separated by the Fabry–Perot resonances of an effective-medium slab. The rules of the correct choice for fully continuous medium slabs are known [200, 201].

Since 2002 scientific groups fabricating MMs have started to apply this algorithm for the characterization of their samples. Then some recent modifications of the method appeared. For example, the authors of [202] noticed that the commonly adopted version of the NRW algorithm (concerning the choice of m) results in a forbidden sign of the imaginary part of one of two EMPs. Since the incorrect sign is prohibited by the passivity principle, in that work one proposed to determine m in a different way—that maintaining the "correct sign" for the imaginary parts at all frequencies. However, it remained unnoticed in [202] that this correct sign does not prevent the material parameters from violating the causality principle. In fact, the Kramers–Kronig conditions not only determine the sign of the imaginary part of a material parameter. Also the real part must vary versus frequency in a physically sound way. Namely, both ε_{eff} and μ_{eff} should grow versus frequency beyond the region of the resonant losses. This obvious requirement was not fulfilled in [202], and the frequency dispersion of the permittivity and permeability retrieved in this work was weird.

As it was already mentioned, violation of the locality in the effective material parameters obtained from (5.46) and (5.47) often was interpreted as a manifestation of the strong SD. Again, let us repeat that it can be a misinterpretation. The strong SD in the medium bulk, even if not seen from the scattering matrix, is always seen from full-wave numerical simulations in the dispersion diagram and in the isofrequency contours calculated for an infinite array. The analysis of the dispersion and isofrequencies for an amount of known MMs for which the antiresonant EMPs were retrieved has shown: At the frequencies of interest the strong SD either is absent in them or arises only in a certain part of the resonance band. In both these cases the correctly retrieved EMP should be local at the frequencies when the strong SD is absent.

In many MMs the strong SD arises at frequencies higher than the resonance band of their m-dipoles. For example, for a lattice of dual SRRs (pairing was needed in order to compensate the bianisotropy) operating in the infrared range that were suggested and studied in work [8] the dispersion diagram does not show any difference from those of magneto-dielectric media. Unlike the lattice of ceramic spheres and like the lattice of OPs considered above the electric and magnetic resonances of these dual SRRs occur below the frequency of the first Bragg resonance. They are continuous media in the frequency region of interest. Refractive index n and characteristic

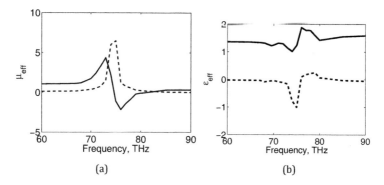

Figure 5.10 Frequency dependencies of the retrieved EMPs of a lattice of infrared SRRs that reproduce the results of [8]: (a)—magnetic permeability (Lorentz's resonance); (b)—dielectric permittivity (antiresonance). The time dependence is $\exp(-i\omega t)$.

impedance Z_c of this lattice retrieved in [8] are both reproduced in Fig. 5.9. These frequency dependencies were correctly retrieved in [8] from R and T of a composite slab using the NRW method. Both refractive index and characteristic impedance do not violate the causality and passivity. In fact, for the refractive index we observe the Lorentzian resonance. For the frequency dispersion of the impedance the Lorentzian dispersion is not obvious. The only implication of causality for the impedance dispersion is the sign of the real part—it must be positive as shown in Fig. 5.9b.

Next, if we assume in accordance with [8] that the composite slab is continuous up to its interfaces and Eqs. (5.45) are correct we obtain the frequency plots for EMPs presented in Fig. 5.10. The plot of the effective permittivity in Fig. 5.9b is a typical example of the antiresonance. The real part noticeably decreases versus frequency below the band of resonant losses (73–80 THz). In the range 73–77 THz the imaginary part has the wrong sign. This antiresonance was misinterpreted as the feature of the strong SD and the band of the resonant losses 73–77 THz was identified with a photonic-crystal stopband (marked in Fig. 5.9). Below we will see that it is not so—the physical meaning of the EMPs can be restored by a correct retrieval.

5.2.5 On So-Called Bloch Lattices

The analysis of the literature data for MMs performed as p-m lattices has shown: On the level of retrieved refractive index and characteristic impedance, the violation of Kramers–Kronig relations even in the resonance frequency range is a rare exception. For the majority of available MMs of this kind these two parameters are fully trustful. However, these two correct parameters tell us nothing about the electric and magnetic polarization responses of the medium. If we retrieve ε_{eff} and μ_{eff} through them we often obtain an antiresonance for one of two EMPs, either for permittivity or for permeability. Why does this antiresonance arise? A general answer is clear—it results from the violation of Maxwell's boundary conditions. A more specific answer is also clear—it results from the wrong identification of n and Z_c with the refractive index and wave impedance of the medium. However, a final answer is needed: What is wrong n or Z_c?

From the theory of continuous media [5] it is known that the surface impedance of a half-space is equal to the wave impedance. However, it is so only if the medium is continuous up to its interface. The violation of Maxwell boundary conditions destroys this equivalence. Formula (5.44) determines the surface impedance of the effective medium normalized to η. So, the mistake is to assume that $Z_c \equiv Z_w$.

In order to understand what the surface impedance of an effectively continuous medium with a discontinuity on the interface is, it is relevant to refer either to solid-state physics or to the theory of photonic crystals (see [37]). In this theory an approximate formula relating the reflection coefficient of a semi-infinite lattice with the so-called *Bloch impedance* is known:

$$R_\infty \approx -\frac{1-Z_B}{1+Z_B} \qquad (5.48)$$

The definition of the dimensionless (normalized to η) Bloch impedance for any orthorhombic photonic crystal is as follows:

$$Z_B = \frac{E^{TA}}{\eta H^{TA}} \equiv \frac{\int_S E(\mathbf{r})dS}{\eta \int_S H(\mathbf{r})dS}. \qquad (5.49)$$

Here E^{TA} and H^{TA} are transversally averaged fields taken at the input or output face of a crystal unit cell. Definition (5.49) applied to a crystal unit cell located on the surface of a half-space or a finite-thickness slab is equivalent to the definition of the normalized surface impedance Z_c. For the infinite lattice the Bloch impedance describes the characteristic impedance of an eigenwave travelling normally to the unit cell face. The concept of the Bloch impedance was borrowed by radio engineers from the solid state physics and physics of photonic crystals and applied to periodically loaded transmission lines (see [204]). For periodically loaded transmission lines the Bloch impedance is defined as the ratio of the voltage $V(0)$ at the input (or at the output—$V(a)$) section of the unit cell to the current $I(0)$ at the input section (or the output current $I(a)$) [204].

The Bloch impedance of a photonic crystal is relevant if and only if the approximate formula (5.48) makes sense. Is it not always so. If there is a noticeable near-field interaction between the crystal planes along the propagation direction formula (5.48) is not adequate and the Bloch impedance is not a relevant parameter. Well, it still can be formally introduced for an infinite lattice, but for a semi-infinite lattice it makes no sense because would be dependent on the depth of a unit cell.

However, in this chapter we concentrate on p-m lattices in which the near-field interaction between the adjacent crystal planes is negligible. These and only these lattices are analogous to transmission lines periodically loaded by lumped impedances. For our lattices the concept of the Bloch impedance can be also borrowed from the theory of crystals. For our MM slabs the retrieved surface impedance is the Bloch one and not the wave impedance Z_w.

Again, the Bloch impedance of an effectively continuous regular MM is not equal to Z_w because Z_w relates the *mean* electric and magnetic fields. The mean field results from the bulk averaging of true fields. The Bloch impedance Z_B relates the *transversely averaged* electric and magnetic fields, fields not averaged along the propagation path z. The mean field entering the definition of Z_w demands an additional averaging. In other words, the Bloch impedance refers to the not yet homogenized lattice, whereas Z_w describes its fully homogenized model.

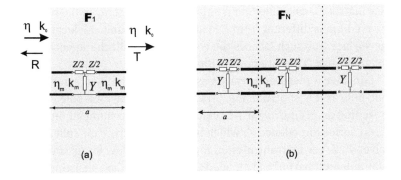

Figure 5.11 A transmission-line model of a Bloch lattice formed by identical p-m grids. The transmission matrix of a transmission line of length a loaded by one four-pole circuit (a) determines the reflection R and transmission T coefficients of a plane wave incident to a single composite layer. It also determines the transmission matrix of an equivalent transmission line comprising N unit cells (b).

Of course, Bloch impedance—an input impedance of an equivalent loaded transmission line—is physically sound and this is why its real part is positive, if it is retrieved correctly. And the NRW method allows us to do it correctly. Transversally averaged fields are continuous across the interface of the MM slab since they are not mean fields but true fields spread only in the tangential plane $(x-y)$. That is why we obtain a correct result for the retrieved value of Z_B at least for a group of regular MMs we call the *Bloch lattices*. As to the value of the retrieved wave impedance Z_w, to assume that $Z_w = Z_B$ is incorrect in the case when Maxwell's boundary conditions do not hold.

Now, let us introduce a transmission matrix F_1 of a unit cell of our lattice, which relates the transversely averaged fields at the input and output of this cell. As it was previously noted, the quantities E^{TA} and H^{TA} contain only one component each. Therefore, we can write

$$\begin{pmatrix} E^{TA}(0) \\ \eta H^{TA}(0) \end{pmatrix} = \overline{\overline{F}}_1 \begin{pmatrix} E^{TA}(a) \\ \eta H^{TA}(a) \end{pmatrix}, \quad \overline{\overline{F}}_1 \equiv \begin{pmatrix} A & B \\ C & D \end{pmatrix}. \quad (5.50)$$

The strong near-field interaction between the lattice unit cells along the propagation direction would make the transmission matrix elements depending on the presence of the neighbors. In this case matrix $\overline{\overline{F}}_1$ calculated for a unit cell located deeply in the bulk of the medium is different from $\overline{\overline{F}}_1$ calculated for a unit cell located on the surface. For such lattices the concept of the Bloch impedance is not applicable. They are not the Bloch lattices.

Consider a piece of a transmission line loaded by a symmetric four-pole as it is shown in Fig. 5.11. Any reciprocal four-pole can be presented as an equivalent T-circuit consisting of a shunt admittance Y and series impedance Z, which in the symmetric case splits onto halves $Z/2$. Let its transmission matrix be $\overline{\overline{F}}_1$. Now let us consider a finite-length periodically loaded transmission line consisting of N such pieces (in Fig. 5.11 $N = 3$). Will its transmission matrix $\overline{\overline{F}}_N$ be equal to $\overline{\overline{F}}_1^N$? For multi-layers of continuous media it is always so [42, 64, 65], because the material parameters of the medium layers (permittivity, permeability and perhaps also the MEC parameter) do not depend on the presence of the adjacent layers. In the periodically loaded lines with ideally lumped loads it is also so. However, there are transmission lines whose real loads have substantial sizes. Substantial loads of two adjacent unit cells of a loaded line can be coupled directly via the mutual inductance or mutual capacitance. Then $\overline{\overline{F}}_N \ne \overline{\overline{F}}_1^N$ and for such loaded transmission lines the Bloch impedance is also not relevant.

Now, we understand which of regular MMs deserves to be called the *Bloch lattice*. It is a lattice for which the equation $\overline{\overline{F}}_N = \overline{\overline{F}}_1^N$ holds with high accuracy, and it behaves as a periodically loaded transmission line with ideally lumped (negligibly small) loads. It is a lattice for which the in-plane near-field interaction of constituents strongly dominates over the near-field interaction between the adjacent crystal planes along the propagation direction. Briefly, it is a lattice for which the Bloch impedance is relevant and for which the NRW retrieval procedure results in the physically sound values n and Z_B, which are both invariant to the number of unit cells N across the effective-medium slab. This is so because the same transmission matrix F_1 of a unit cell is retrieved for any $N = 1, 2, 3 \ldots$ from corresponding R and T.

If it is really so for any propagation direction, our first-principle dynamic homogenization is applicable for all eigenwaves. If it holds only for a narrow sheer of directions around the z-axis, our model is applicable only for eigenwaves propagating inside this cone. Another restriction—the concept of the Bloch impedance makes sense only at frequencies below the Bragg reflecting resonance $\min(a, b) = \lambda/2$, where λ is the wavelength in the matrix (perhaps modified by the static polarization of particles). At higher frequencies formula (5.48) obviously becomes inadequate because at high frequencies the impact of polaritons grows (also, when $\min(a, b) > \lambda$ the high-order Floquet spatial harmonics become propagating in the lattice together with the fundamental one).

5.2.6 A Correct Retrieval Procedure for Bloch's Metamaterial Lattices

The concept of the Bloch lattice allows one to apply the existing theory of periodically loaded transmission lines. Recall that the Bloch impedance is a parameter of the original (periodically loaded) line, whereas the wave impedance Z_w is the parameter of a homogenized line (in which the original lumped loads are spread over the unit cell).

Comparing the known expression for the impedance Z_w of a homogenized periodically loaded line [204] with the expression for Z_B in the original periodically loaded line [205], one may obtain:

$$\frac{Z_w}{\eta} = \frac{Z_B + \Gamma}{(k_m/k_0)^2 + Z_B\Gamma}, \quad \Gamma \equiv \frac{(k_m/k_0)^2 - j(q/k_0)\tan(k_m a/2)}{(q/k_0) - j\tan(k_m a/2)}. \tag{5.51}$$

Here η is the wave impedance of the transmission line connecting the periodically inserted loads and $k_m = k_0 n$ is the wave number in this host line. For the original p-m lattice the wave impedance Z_w is the ratio of the bulk averaged fields, whereas the Bloch impedance is that of the transversely averaged fields. The analogy with an electromagnetic field gives $n = \sqrt{\varepsilon_m}$ and in this case η in (5.51) is the wave impedance of the host medium.

A more difficult derivation of (5.51) can be found in works [14, 15]. It does not use the analogy with transmission lines and is based on the consideration of the wave reflection by a monolayer illuminated by a normally incident plane wave. The monolayer is a MM slab with $N = 1$ unit cell across it. The grid of p- and m-dipoles is located at the central plane of the slab of thickness a filled with the host medium of permittivity ε_m. The planar homogenization replaces the p- and m-dipoles of the grid by an effective sheet containing both electric surface current $J_e = j\omega p/S$ and magnetic surface current $J_m = j\omega m/S$. Solving this problem one may express the ratio $\eta J_e/J_m = (\eta p/m) \equiv \gamma$ via the surface impedance Z_1 of the monolayer[d]. This surface impedance is related with the transmission matrix $\overline{\overline{F}}_1$ of the unit cell. This matrix is the same for an isolated monolayer and for a unit cell of a stack of monolayers—the original Bloch lattice. The Bloch impedance Z_B can be expressed through the elements of $\overline{\overline{F}}_1$. This way the Bloch impedance of an infinite lattice is related to the surface impedance of a monolayer. Since Z_1 has been already expressed via the coefficient γ, one can obtain the relationship between the coefficient γ and Z_B:

$$\gamma \equiv \frac{\eta J_e}{J_m} = \frac{\sqrt{\varepsilon_m} - jZ_B \tan(k_m a/2)}{Z_B - j\sqrt{\varepsilon_m}\tan(k_m a/2)}.$$

Substituting this relation into Eq. (5.39) for parameter γ one derives (5.51) alternatively—without involving the theory of loaded transmission lines. Notice that formula (5.51) refers only to the lattice of combined p-m grids. In [16] a relationship between Z_B and Z_w similar to (5.51) was derived for the case of alternating p- and m-grids.

Expressing the wave impedance through the Bloch impedance in accordance with (5.51), we modify the NRW retrieval procedure so that to extract the EMPs corresponding to our dynamic homogenization model of an infinite lattice. Although these parameters are also retrieved from the scattering matrix of a slab (like antiresonant EMPs), they have clear physical meaning because are equivalent to those defined by formulas (5.19) and describe the electric and the magnetic responses of the medium unit cell. This is so because in the infinite and in the finite Bloch lattice the transmission matrix of a unit cell is the same, and, therefore, the refractive index and Bloch impedance are the same.

As to the antiresonant EMPs shown in Fig. 5.10, they are meaningless—that effective permittivity does not describe the

---

[d]Do not mix it with the magnetic grid impedance Z, the last one is only a series load in the unit cell center.

electric response of the medium and that effective permeability does not describe the magnetic response.

5.2.6.1 Locality of retrieved parameters

Using the obtained relationship between the Bloch impedance and wave impedance of the effective bulk medium, one may correctly extract the bulk EMPs from the reflection $R^{(N)}$ and transmission $T^{(N)}$ coefficients of a MM slab with N unit cells across. No need to involve Drude transition layers for this correct retrieval. Of course it is possible if and only if the MM is a Bloch lattice.

As we have already clarified, the Bloch impedance does not correspond to the homogenized model, it is characteristic for a discrete set of crystal planes. Using the description of the lattice in terms of Z_B we may be quiet about Maxwell boundary conditions: they are respected because applied to transversally averaged microscopic fields and not to mean fields.

Thus, first, we find n and Z_B from $R^{(N)}$ and $T^{(N)}$ and, second, we find the bulk impedance Z_w from (5.51) and finally retrieve the bulk parameters as simply as

$$\varepsilon = n \frac{\eta}{Z_w}, \quad \mu = n \frac{Z_w}{\eta}. \tag{5.52}$$

It is important to note that we have no problem with the expression of retrieved n via the effective permittivity and permeability. The refractive index is a fully macroscopic parameter. For a periodically loaded line n is common for two its descriptions—via the Bloch eigenmode propagating in the original line and via the wave in the homogenized one. Similarly, for a p-m lattice n is common for two descriptions of the eigenmode—via the transversal averaging of the true fields and polarizations representing the lattice as a set of polarized sheets and via the complete (bulk) averaging representing the lattice as a continuous medium. The problem arises only for the impedances.

Let us take n and Z_B of the lattice of infrared SRRs retrieved in work [8] and depicted in Fig. 5.9 as the input data substituted into (5.51) and (5.52). The result is shown in Fig. 5.12. The plot for the effective permeability has changed though not dramatically. However, the effective permittivity has changed drastically. Now, it

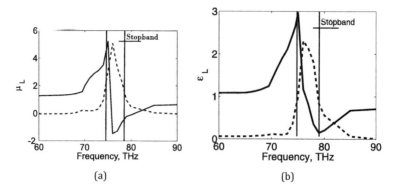

Figure 5.12 Frequency dependencies of the correctly retrieved EMPs of a lattice of infrared SRRs replace the antiresonant EMPs depicted in Fig. 5.10: (a) Effective permeability. (b) Effective permittivity. Index L means that the material parameter experiences the Lorentz's resonance. Time dependence is $\exp(-i\omega t)$.

does not manifest the antiresonance any more, it is a Lorentzian function.

We note that the exact numerical investigations of these lattices showed that the transmission matrix $\overline{\overline{F}}_N$ of the slab with N unit cells across it is with very high accuracy equal to $\overline{\overline{F}}_1^N$, i.e., the lattice is the Bloch one. Therefore, there is no need to specially prove that the retrieved EMPs shown in Fig. 5.12 are equal to those of an infinite lattice. The transmission matrix $\overline{\overline{F}}_1$ of a monolayer contains all the information needed to find both the refractive index n (or wave number q) and the wave impedance Z_w of the infinite lattice eigenmode. Parameters Y and Z of the lattice crystal plane are the same for the isolated monolayer. Together with the unit cell length a and the host medium permittivity ε_m they completely determine the eigenmode of the lattice. Therefore, the material parameters correctly retrieved from the reflection and transmission coefficients are namely those of the infinite Bloch lattice.

So, parameters n and Z_B depicted in Fig. 5.9 properly describe the lattice eigenmode, whereas the Bloch impedance is not the effective-medium wave impedance Z_w. Parameters ε_L and μ_L depicted in Fig. 5.12 properly describe the magnetic and electric polarizations of the effective medium, since are retrieved through Z_w. In Figs. 5.9 and 5.12, the locality of EMPs holds in the whole band of resonant losses 73-79 THz previously treated as a photonic-crystal stopband.

Within the framework of the ECONAM project (see in Preface) the author of this book has revised several regular MMs, for which the antiresonant EMPs were retrieved via n and Z_B treated as Z_w. One may share two types of obtained numerical results. The results of the first type corresponded to the failure of the dynamic homogenization model, when the correctly retrieved EMPs still manifest the non-local frequency behavior. In each of these cases the values n and Z_B retrieved from $R^{(N)}$ and $T^{(N)}$ depended on N. This meant that the MM under study was not the Bloch lattice. Namely, this result was obtained for the so-called *optical fishnet MM* [206], for a lattice of densely packed plasmonic SRRs [207], and for a lattice of plasmonic oligomers [208]. For these three types of nanostructured MMs the Bloch impedance makes no sense, and our homogenization model is not applicable.

Besides a cubic lattice of the dimers of infrared SRRs, a cubic lattice of those of the dimers of OPs (operating at microwaves) were analyzed, as well as a one-phase cubic lattice of the Mie-resonant ceramic spheres (microwaves), a square lattice of infinitely long ceramic cylinders (microwaves), and orthorhombic lattices of silicon spheres (visible range). For these p-m lattices the NRW retrieval procedure gave the same n and Z_B for different number N of unit cells across the slab, i.e., these regular MMs were the Bloch lattices. For each of these MMs, the frequency curves for correctly retrieved EMPs represented the typical pattern of the Lorentzian resonance, whereas the usual NRW retrieval procedure resulted in the antiresonance of either permittivity or permeability. This analysis occupied a rather long period of time (2006–2010) and resulted in a conclusion that the first-principle homogenization model developed above is adequate for all Bloch's lattices at the frequencies of their effective continuity [6, 194]. We will see below that this conclusion was a bit hasty: Low-loss strongly resonant Bloch's lattices at the edges of their resonant stopband can be effectively continuous, whereas our model predicts for them the strong SD.

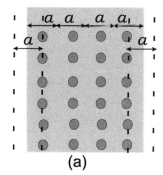 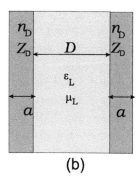

Figure 5.13 Introduction of two transition layers for a finite-thickness lattice of resonant p-m particles in a dielectric matrix with a thickness of $D_s = Na$ and permittivity ε_m: (a) How the transition layers are located. (b) The sectionally homogeneous structure equivalent to the initial slab.

5.2.7 Retrieval of the Transition Layer Parameters

The introduction of the transition layers for a MM with the bulk effective material parameters ε_L and μ_L is the key point of the theory. Since for mean fields our MM slab has no homogeneous interfaces, a transition layer mimics the surface jump of the mean field tangential components. This transition layer should play an equally important role for an effective-medium sample with $N = 400$ and for a slab with $N = 4$. In both cases, the variation of the tangential mean field across two transition layers significantly affects the coefficients R and T of a homogenized slab allowing these coefficients to be equal to R and T of the original structure. An original slab of thickness $D = Na$ and a homogenized slab of thickness $D = (N + 1)a$ with a three-layer structure are illustrated by Fig. 5.13.

In the notations of the effective parameters we will use the index D in the honor of P. P. Drude and the index L in the honor of H. A. Lorentz. Let the parameters ε_L and μ_L be already found for this lattice from the coefficients $R \equiv R^{(N)}$ and $T \equiv T^{(N)}$ as we have done above. Since R and T can be expressed via the EMPs of this sectionally homogeneous structure, and the parameters of the central layer are known one can determine the missing parameters of transition layers from the same R and T.

It is not surprising that we use twice the same scattering parameters of the slab (in the first case, we used these quantities in order to find Z_B and n, and in the second case for finding the parameters of the transition layer). Both reflection and transmission coefficient refer to the mean field calculated in free space and equally can be related either to the transversally averaged fields or to the bulk averaged fields at the slab interface. In free space they are equivalent.

Recall: When we relate R and T to transversally averaged fields we do not need any transition layers. The composite slab is in this model a stack of N unit layers or, equivalently, a stack of unit cells of a periodically loaded transmission line. If we relate R and T to mean fields in the homogenization model we have to represent a composite slab as a three-layer structure shown in Fig. 5.13. There are only two unknown parameters—those of two identical transition layers. They can be successfully expressed through the same R and T.

The transmission matrix of the whole slab has only two independent components A and B—for a periodically loaded transmission line of symmetric unit cells we have $D = A$ and $C = (1 - A^2)/B$ [204]. It is easy to express these two independent components of the transmission matrix via R and T:

$$A = \frac{1 \pm \sqrt{2(R^2 + T^2) - 1}}{2T}, \quad B = \frac{1 - R}{T} - A. \tag{5.53}$$

Matrix $\overline{\overline{F}}_N$ of the whole slab

$$\overline{\overline{F}}_N \equiv \begin{pmatrix} A & B \\ C & D \end{pmatrix}.$$

in our model is the product of the transmission matrices of the central layer $\overline{\overline{F}}_{CL}$ and of the transition layers $\overline{\overline{F}}_{DL}$. Namely, in is equal to (see more details in [17]):

$$\overline{\overline{F}}_N = \overline{\overline{F}}_{DL} \overline{\overline{F}}_{CL} \overline{\overline{F}}_{DL},$$

$$\overline{\overline{F}}_{DL} = \begin{pmatrix} \cos(k_0 a n_D) & -i Z_D \sin(k_0 a n_D) \\ \frac{i \sin(k_0 a n_D)}{Z_D} & \cos(k_0 a n_D) \end{pmatrix},$$

$$\overline{\overline{F}}_{CL} = \begin{pmatrix} \cos q D & -i Z_L \sin q D \\ i \frac{\sin q D}{Z_L} & \cos q D \end{pmatrix}. \tag{5.54}$$

Here, $q \equiv k_0 n$ and $Z_L \equiv Z_w$ are the wave number of the central layer and its wave impedance, respectively. The index L is chosen in the honor of H. A. Lorentz. It stresses that the whole procedure makes sense if and only if the EMPs of the central layer are Lorentzian. In (5.54) n_D and Z_D are the refractive index and the wave impedance of the transition layers, respectively. The index D is again introduced in the honor of P. P. Drude (though the concept of Drude's transition layers was different).

We do not try to describe the transition layers by a set of bulk EMPs because such material parameters will be not physically sound. In fact, our transition layers should not be those of any feasible bulk material. They simply mimic the jumps of bulk mean fields at the MM interface. It is even not evident that their wave impedance Z_D and their refractive index n_D in our retrieval procedure will be compatible with the causality and passivity requirement. Meanwhile, it is methodologically necessary to have the description of a transition layer via some physically sound parameters. If n_D and Z_D are not physically sound, e.g., n_D decreases versus frequency in the lossless frequency region and/or Z_D has the negative real part, we will consider our homogenization model as failed.

Numerous calculations by the author have shown that for Bloch lattices parameters n_D and Z_D are physically sound and description of the transition layer as a piece of a transmission line is adequate if the transition layers are located as it is shown in Fig. 5.13. The interfaces between the transition and central layers correspond to the edge (left and right) crystal planes of the MM slab. The thickness D of the central layer in our model equals $D_s - a = a(N - 1)$—then by one period than the physical thickness of the composite slab. For cubic lattices the correct thickness of the transition layer (offering physically sound n_D and Z_D) is equal to the lattice period a, and the layered effective-medium slab is thicker than the original composite slab by one period a.

For non-cubic (orthorhombic) Bloch lattices the proper thickness of the transition layer (resulting in physically sound n_D and Z_D) is in between maximal and minimal lattice periods. Analytical relations for this case were not derived. The correct choice of the thickness for a non-cubic lattice required a numerical fitting.

In this book we restrict our consideration by the case of a cubic lattice. Since the thickness of the transition layer a is optically small, its transmission matrix can be simplified as

$$\overline{\overline{F}}_{DL} = \begin{pmatrix} 1 & -iZ_D n_D k_0 a \\ i(n_D/Z_D)k_0 a & 1 \end{pmatrix}.$$

From (5.54) we have the following expressions for the components of the transmission matrix $\overline{\overline{F}}_N$ of the whole slab:

$$A = \cos qD + k_0 a n_D \left[\left(\frac{Z_D}{Z_L} + \frac{Z_L}{Z_D} \right) \sin qD + k_0 a n_D \right], \quad (5.55)$$

$$\frac{iB}{Z_L} = \sin qD + k_0 a n_D \left(\frac{2Z_D}{Z_L} \cos qD + k_0 a n_D \sin qD \right). \quad (5.56)$$

The refractive index and the wave impedance of the transition layers are found as a solution of the system of Eqs. (5.55) and (5.56):

$$n_D = \frac{1}{k_0 a} \sqrt{G \left(\frac{A - \cos qD}{\sin qD} - G \right)}, \quad Z_D = \frac{Z_L G}{k_0 a n_D}, \quad (5.57)$$

where it is denoted

$$G = \frac{(jB/Z_L) - \sin qD}{\cos qD}.$$

Parameters A and B have already been expressed through R and T in Eqs. (5.53) and are known.

Figure 5.14 shows the result for a finite-thickness lattice of infrared SRRs in the case $N = 10$. It can be seen in the plots of n_D and Z_D that the resonance range of the bulk effective material parameters of the MM corresponds to the resonance of the wave impedance of the transition layer. The refractive index n_D does not experience a resonance and grows versus frequency. Both frequency dependencies of Z_D and n_D and the signs of their real and imaginary parts do not violate the locality conditions. It is clear that the implementation of the locality conditions for Z_D and n_D means the absence of the SD in the a-long pieces of an effective transmission line describing the transition layers.

In contrast to the EMPs of the central layer, the parameters of the transition layers are mesoscopic—depending on the thickness of the

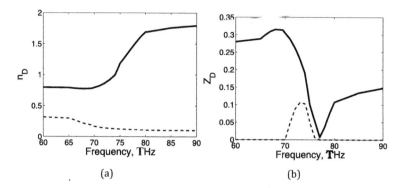

Figure 5.14 Parameters of transition layers for the lattice of infrared SRRs with thickness of $N = 10$ cells: (a) refractive index and (b) wave impedance. Time dependence is $\exp(-i\omega t)$.

MM slab. This is evident from comparison of Figs. 5.14 and 5.15. The plots in the latter figure correspond to $N = 20$. The fact that the parameters of transition layers depend on N is not surprising. It is immediately seen from (5.55) and (5.56) that n_D and Z_D depend on D due to the presence of terms $\sin qD$ and $\cos qD$ in the elements of the structure transmission matrix. The mesoscopy of the transition layers does not harm to the model applicability. It is more important that the same transition layer can be attributed to both illuminated and rear interfaces of the composite slab.

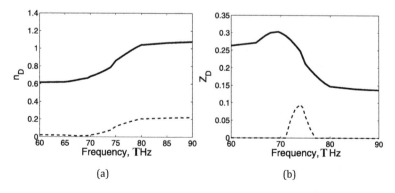

Figure 5.15 Parameters of transition layers for the SRR lattice with thickness of $N = 20$ cells: (a) refractive index and (b) wave impedance.

For the other studied Bloch MMs (see their list in the end of the previous subsection) the results of this retrieval were similar. So, the non-locality of the surface of a resonant composite slab if this composite is a Bloch lattice is successfully described in a condensed form by the introduction of thin transition layers at both interfaces of the slab. These layers are optically thin and described by two parameters n_D and Z_D.

It is important to note that, at $N \le 3$, the requirement of the locality of the transition layers in all studied cases was not implemented. This refers not only to the lattice of SRRs. For all studied Bloch's lattices the refractive index n_D experienced the antiresonance if $N \le 3$. Most likely, this result does not mean that the transition layers are not suitable. On the contrary, it indicates a minimal thickness of the MM slab that can be described by the bulk EMPs. If $N \le 3$ it is possible to describe the slab with resonant artificial magnetism via physically sound bulk EMPs and transition layers. However, this description is physically not adequate because such a slab is a metasurface rather than a volumetric slab. Physically meaningless parameters of its transition layers are indicators of this inadequacy.

5.3 Recent Progress

5.3.1 Effective Surface Sheets Instead of Transition Layers

As it was already mentioned, the finite-thickness transition layers can be successfully squeezed to sheets of the electric and magnetic surface currents. In other words, the jumps in the volumetrically averaged tangential electric and magnetic fields on either side of the interface between a MM and air are mimicked by equivalent metasurfaces located at the interfaces [21, 22]. Metasurface can be defined as a surface arrangement of optically small resonant scatterers. In other words, metasurface is a 2D MM, a planar analogue of a volumetric MM.

Nowadays, a commonly recognized homogenization model of a metasurface is based on so-called *generalized sheet transition conditions* (GSTCs), often attributed to C. Holloway and E. Kuester, but in fact introduced earlier—by Thomas Senior, John Volakis, Mithat Idemen, and Hamit Serbest (see in [209]). In work [20] the equivalence of the model based on GSTCs to other known models of a metasurface was proved.[e] Therefore, below we concentrate on this model.

GSTCs are equations containing in the left-hand side the jumps of the tangential components of the mean electric and magnetic fields. In the standard model of the metasurface the true fields are transversally averaged and called mean fields [210]. However, as it is explained in [209] this approach is valid only for metasurfaces located in a homogeneous matrix, e.g., in free space. If a metasurface (real or fictitious as in the present situation) is located on an interface of two continuous media, GSTCs hold if and only if the true field is averaged not only transversely but also over a certain interval in the z-direction. This means that the GSTCs in the new homogenization model of a MM slab are introduced for bulk mean fields.

There are several equivalent forms of GSTCs for effectively homogeneous metasurfaces. GSTCs may contain as parameters the electric and magnetic surface susceptibilities of the effective metasurface. However, such a form is rather misleading because the homogenization model in this case would deal with both transversally averaged (still microscopic) and fully macroscopic mean fields. The most convenient and general form of GSTCs comprising namely the mean fields combined with surface polarizations was derived in [209]:

$$\mathbf{E}_t^+ - \mathbf{E}_t^- = j\omega \mathbf{n} \times \mathbf{M}^{(s)}{}_t - \nabla_t \frac{P^{(s)}{}_n}{\epsilon},$$

$$\mathbf{n} \times \mathbf{H}_t^+ - \mathbf{n} \times \mathbf{H}_t^- = j\omega \mathbf{P}^{(s)}{}_t + \nabla_t \times \mathbf{n} \frac{M^{(s)}{}_n}{\mu}, \quad (5.58)$$

where ϵ and μ are EMPs of the substrate—medium at which the metasurface is located. $\mathbf{P}^{(s)}$ is the electric surface polarization, $\mathbf{M}^{(s)}$ is the magnetic one, indices "n" and "t", respectively, denote the normal

---

[e]There are two other relevant models of resonant optically thin and dense grids—that of the impedance matrix, also called the *circuit model* and that of the so-called *collective polarizabilities*.

and tangential components. Effective parameters of the metasurface may relate $\mathbf{P}^{(s)}$ and $\mathbf{M}^{(s)}$ with the transversally averaged fields [210]. In work [21] explicit formulas relating these effective parameters, called *surface susceptibilities* with the reflection R and transmission T coefficients of a finite-thickness slab and with its bulk permittivity and permeability. In this work one considered a MM slab as an effectively continuous layer whose interfaces are covered by two identical metasurfaces.

The electric and magnetic surface susceptibilities ξ_E and ξ_M relate $\mathbf{P}^{(s)}$ and $\mathbf{M}^{(s)}$ with the transversally averaged electric and magnetic fields taken at the metasurface plane. They were found in [21] together with ε and μ from the exact numerical simulations of two samples of the same MM. The first sample was a slab containing N monolayers and the second one contained $2N$ monolayers. In this model it is assumed that the electric and magnetic surface susceptibilities of the transition metasurfaces are not mesoscopic, they keep the same for two different thicknesses of the MM slab. Two sets of R and T allow one to find four scalar parameters ε, μ, ξ_E and ξ_M rather easily. However, the plots of the obtained surface susceptibilities turned out to be not compatible with the idea of the local response. Both real and imaginary parts of these surface susceptibilities oscillate versus frequency changing the sign. This behavior is prohibited for physically realized metasurfaces. As to the retrieved bulk EMPs, everything is OK for them: They nicely exhibit the Lorentzian resonances and are equivalent to the EMPs of the infinite lattice.

This result is a clear indication to the fitting nature of these effective metasurfaces. Is it a drawback of the whole model or is it so because the possible mesoscopy of the effective metasurface was neglected (the same surface parameters were attributed to the composite slabs of different thicknesses)?

In [22] it is assumed that ξ_E and ξ_M can be mesoscopic: For slabs with different thickness ξ_E and ξ_M are different. However, the retrieval of ξ_E and ξ_M from the same pair of R and T, as it was done above for the bulk transition layers, turned out to be impossible because the surface electric and magnetic polarizations entering the GSTCs respond not to the mean field but to transversally averaged field. Therefore, only bulk EMPs were retrieved in [22]—the surface susceptibilities could not be found.

It seems that to retrieve these mesoscopic surface polarizabilities is not so simple because the same model should deal with both bulk mean fields and transversely averaged fields. Above, we have already mentioned that there is an alternative description of a metasurface response relating the electric and magnetic surface polarizations with the electric field of the *incident* plane wave. Such parameters are called *collective polarizabilities* of the metasurface (see formulas (12) of work [209]). Adding these parameters to the bulk EMPs we will avoid the mess in our homogenization model. Using this approach instead of that of [21, 22] we avoid mixing up the bulk mean fields and transversely averaged fields. The retrieval of the collective polarizabilities combined with the retrieval of the bulk parameters from the same pair of R and T seems to be realistic. Such a study has been not done, yet, and it is a new field for enthusiastic researchers.

5.3.2 Bianisotropic Parameters of Non-Bianisotropic Materials

5.3.2.1 Resonance band: optical losses versus Bragg resonance

As it was already mentioned, the effectively continuous Bloch lattices of non-bianisotropic p- and m-dipole scatterers can be described by Lorentzian permittivity and permeability after a proper modification of the NRW retrieval. However, as we have also mentioned, the effective continuity of the Bloch lattice can be broken at some frequencies within the resonance band, where a strong SD arises. Then both our first-principle dynamic homogenization model and our retrieval method based on this model fail. The retrieved EMPs of an effective slab and the EMPs of an infinite lattice calculated directly still coincide because the transmission matrix $\overline{\overline{F}}_1$ is the same, but the frequency dispersion of these parameters is wrong and violates the Kramers–Kronig relations. This result was obtained for lossless lattices, for example, if we replace the OPs of copper and SRRs of silver (see above) by the same particles performed of perfectly conducting metal. In this case the band of resonant losses is replaced by the resonant stopband in which the effective wavelength shortens so that the Bragg regime is implemented. Therefore, the antiresonant EMPs for this case were treated by the author of this book as the indications that these Bloch lattices were not effectively continuous at the frequencies of the resonance band.

However, in 2011 Andrea Alù revisited the issue of low-loss Bloch's lattices and revealed that at the edges of the resonant stopband there are two frequency intervals for which the first-principle homogenization model developed above gives the antiresonant frequency behavior for EMPs. The effective medium at these frequencies is not effectively discrete. Therefore, it can be described by local EMPs if the homogenization model is properly modified.[f]

To understand why our homogenization model can fail for an optically dense Bloch lattice let us first compare the resonance of an effectively continuous MM with the lattice resonance of a photonic crystal that represents in the same MM at higher frequencies.

There is a major difference between the Bragg-like regime when in a lossy resonant MM the lattice period a tends to $\lambda_{\text{eff}}/2$ due to the resonance wave shortening, and the truly Bragg regime when λ_{eff} is close to the wavelength in the host medium and is affected by the lossy lattice only slightly. In the second case we obviously observe the lattice Bragg resonance when the propagation is prohibited due to the destructive interference of plane waves radiated by the polarized crystal planes in the forward direction. Simultaneously, the waves partially reflected by the crystal planes backward interfere constructively. This regime holds in the Bragg stopband and at its edges the medium is fully transparent.

In the first case the Bragg phenomenon as such is not observed and the lattice remains an effectively continuous medium with the Lorentzian frequency dispersion of the electromagnetic response. In fact, at the resonance the eigenmode decay increases drastically and the interference of partial waves produced by crystal planes is

---

[f]Notice that A. Alù on that time was not aware of the corresponding works of the present book author. He referred to the works by Mario Silveirinha, whose homogenization model gave the same results (see the story in Preface).

suppressed. They weakly interfere because of different amplitudes. The eigenmode propagation is obstructed and may be even fully prohibited in this frequency range. However, it occurs not due to the Bragg phenomenon but due to resonant losses obviously accompanying the anomalous dispersion.

What is the reason for the suppressed propagation: the Bragg regime or the regime of the resonant losses? The answer is different for different frequencies. The polarizabilities of the small constitutive particles obviously experience the Lorentzian resonance. Exactly at the resonance frequency the real parts of these polarizabilities vanish and the imaginary parts attain the maximum. The dissipation dominates and the propagation is forbidden due to resonant losses.

However, aside of the resonance frequency there is a contest of two processes. The first process is the resonance wave shortening, favorable for the interference of the partial waves. The second process is the resonant dissipation that may be high enough to suppress the interference of the partial waves. The result of this contest is different for different MMs. And for the same MM it can be different at different frequencies of the resonance band.

In one of two numerical examples of doubly negative MMs considered above the medium has rather substantial optical losses. This is the example of Omega-dimers. These losses arise from the finite conductivity of copper and prevent the Bragg resonance at 3.83–3.94 GHz. The medium is opaque due to high attenuation at 3.83–3.94 GHz. In the backward-wave regime that holds at 3.94–4.21 GHz optical losses are still substantial: The negative refractive index is of the order of $n = (-1 \cdots -3) \cdot (1 - j0.01)$. Aside of the resonant stopband, e.g., in the backward wave frequency region the medium is continuous and our model does not fail.

In another example—a two-phase lattice of ceramic spheres—the medium had low optical losses because the ceramics was assumed to be of high quality. And within the Mie resonance band of our spheres we observed the Bragg resonance. In the backward-wave region the negative refractive index of the medium was of the order of $n = (-1 \cdots -3) \cdot (1 - j10^{-4})$, i.e., optical losses were really very small. However, we remember that this medium was effectively discrete in this band.[g]

These two examples show that lossy MMs are more favorable for homogenization that lossless MMs. If optical losses of the constituents are absent or very low, the resonant stopband is the truly Bragg stopband that arises around the resonance frequency. In this band the MM is a photonic crystal without local EMPs. However, at the edges of this stopband, at frequencies where the medium is transparent, it is effectively continuous and is not a photonic crystal anymore. This was understood in work [24]. For an infinite array of equivalent spheres with very high purely real permittivity the dynamic EMPs found from Eqs. (5.19) manifest the antiresonant behavior. This is the indication of the strong SD. However, the SD is strong only within the resonant stopband. Outside it the SD is weak!

The author of the present book considered the cases when his model predicted the antiresonant EMPs as the indications of the strong SD in the whole resonance band—at frequencies where the medium was opaque and at frequencies where it was semitransparent. For example, the set of antiresonant EMPs was obtained in [211] for a lattice of core-shell spheres, experiencing both electric and magnetic resonances. These core-shell sphere in [211] consisted of a submicron silica core and tiny silver nanoshell. Plasmonic resonance of the nanoshell intersects with the magnetic Mie resonance of the core. The aim of that work was to obtain a doubly negative MM though in this design solution the backward-wave regime is impossible without strong SD [197].

Nevertheless, the author of [211] has calculated the permittivity and permeability of his self-resonant photonic crystal as if it was an effectively continuous medium and his EMPs manifested an antiresonance, whereas the resonance band was a truly Bragg stopband. Optical losses in work [211] were practically neglected (the complex permittivity of silver was idealized using the Drude dispersion model unsuitable for the frequency range where the structure should have operated). Moreover, the backward-wave

---

[g]Let us repeat: Any doubly negative medium implemented of small resonant scatterers is either too lossy or too spatially dispersive for imaging purposes.

regime was obtained (for a specific propagation direction) at frequencies higher than the resonant Bragg stopband and the last one was higher than the fundamental Bragg stopband of the lattice. Definitely, aside the resonant stopband such an effective medium though transparent was strongly spatially dispersive. However, this was not the case of the lattice studied in [24].

5.3.2.2 An effectively continuous resonant lattice without optical losses

Our further consideration concerns only a cubic lattice of simple dielectric spheres. Following to work [24] we assume that the permittivity of the spheres is equal to $\varepsilon_s = 120$ and the diameter of a sphere normalized to the lattice period is equal $D/a = 0.45$. This lattice has a quite dense arrangement—the gap between two adjacent spheres is equal 10% of the lattice period a. The p-m dipole model of the response is adequate for such spheres in the range of normalized frequencies $k_0 a \leq 1$. The host material in these calculations is free space.

In Fig. 5.16(a) EMPs of this lattice versus normalized frequency $k_0 a$ are calculated using our first-principle homogenization model. At $0.6 < k_0 a < 0.65$ the resonant stopband related to the m-dipole Mie resonance of spheres is located. Within this band is spite of the absence of losses EMPs become complex and lose their physical meaning. This part of the resonance band corresponds to the Bragg resonance. Therefore, we do not show here EMPs calculated for normalized frequencies $k_0 a = [0.6 - 0.65]$.

Beyond this band there are two frequency regions $k_0 a = [0.5-0.6]$ and $k_0 a = [0.65-0.7]$ where λ_{eff} is though larger than $2a$ but is rather close to this threshold. In these regions we observe the antiresonant behavior of our dynamic permittivity. This can be interpreted as the evidence of the strong SD. One may think that over the whole band of the magnetic Mie resonance—from $k_0 a = 0.5$ to $k_0 a = 0.7$ the effective medium is not continuous. However, in work [24] it was shown that the SD is strong only within the Bragg stopband $k_0 a = [0.6-0.65]$ and in the bands $k_0 a = [0.5-0.6]$ and $k_0 a = [0.65-0.7]$ the SD in the medium is weak, i.e., the medium is effectively continuous. Simply, our dynamic homogenization model

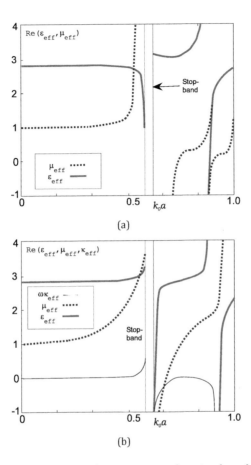

Figure 5.16 Effective material parameters of a simple cubic lattice of lossless dielectric spheres with $\varepsilon_s = 120$ and normalized diameter $D/a = 0.45$ calculated in accordance with [24]: (a) EMPs resulting from formulas (5.19) and called in [24] "equivalent" parameters; (b) three EMPs calculated in accordance with formulas (5.65) and (5.66).

for these bands is not applicable, it gives a wrong prediction and must be revised. A more advanced homogenization model was suggested in [24, 25] that allows the physical meaning of EMPs to be restored in these two frequency bands.

First, recall that the definitions (5.17) and (5.18) leave for both ε and μ the possibility to indirectly depend on the wave vector $\mathbf{q}$.

In fact, we combine in our model two formulas for the mean p-polarization **P** expressed through both local and mean fields. In the first expression $\mathbf{P} = \mathbf{p}/V = (a_{ee}/V)\mathbf{E}^{loc}$ the mean polarization is that of a reference dipole divided by the cell volume, in the second expression $\mathbf{P} = \varepsilon_0(\varepsilon - \varepsilon_m)\mathbf{E}$ it is the medium unit cell polarization. If and only if the relation between $\mathbf{E}^{loc}$ and **E** does not depend on the wave vector **q**, our model delivers the local ε independent on **q**. However, we cannot be sure that in a resonant medium with WSD $\mathbf{E}^{loc}$ and **E** are related by the CMLL formulas even in the bulk!

Earlier we considered the unambiguous relation between $\mathbf{E}^{loc}$ and **E** as the prerequisite of the effective continuity. However, what if the wave vector **q** enters these relations as a small correction term? Is it a weak spatial dispersion or a strong one? Probably, the case when the local and mean electric fields are related with one another via a term weakly (e.g., linearly) depending on **q** the effective medium is still continuous whereas our first-principle model gives the non-local effective permittivity. In other words, our EMPs may still depend on **q** even if the SD is weak in the effective medium. If so, our homogenization model gives wrong predictions.

If in the two questionable frequency bands located at both sides of the resonant stopband the effective permittivity depends on **q**, it implies its non-locality and for a given propagation direction its frequency behavior not compatible with causality is possible. To get rid of this non-physical artifact, we have to take into account that our effective permittivity depends on **q**. Since the medium is effectively continuous this dependence can only be sufficiently weak and can be approximated by a linear functions of **q**. As we have already seen in Chapter 3, the linear q-dependence can be shared out from the non-local permittivity in the form of the effective bianisotropy (together with the redefinition of the auxiliary field vectors). This transformation was done in [25]. Alternatively, one may revise our homogenization model defining the medium electric and magnetic polarization as those linearly depending on **q** [24]:

$$\mathbf{P} = \varepsilon_0(\varepsilon_{\text{eff}} - \varepsilon_m)\mathbf{E} - \kappa^o_{\text{eff}}\mathbf{q} \times \mathbf{H}, \qquad (5.59)$$

$$\mathbf{M} = \mu_0(\mu_{\text{eff}} - 1)\mathbf{H} + \kappa^o_{\text{eff}}\mathbf{q} \times \mathbf{E}. \qquad (5.60)$$

Here, the upper index o implies that the MEC parameter κ^o_{eff} is introduced for an ordinary p-m lattice—that with local response described by non-bianisotropic relations (5.24) and (5.25). For lossless resonant lattices the fictitious MEC parameter κ^o_{eff} turns out to be significant in two frequency bands located around the resonant Bragg stopband. If in these (previously antiresonant) bands the medium is transparent the new definition of the medium susceptibilities given by relations (5.59) and (5.60) is successful—it grants a set of local EMPs. Old effective parameters ε and μ delivered by the first-principle homogenization model were called in works [24, 25] "equivalent" material parameters.

The model was extended to the case when the constitutive particles are bianisotropic such as OPs. In this case an "extraordinary" MEC parameter κ^e_{eff} is added and definitions (5.59), (5.60) generalize as

$$\mathbf{P} = \varepsilon_0(\varepsilon_{\text{eff}} - \varepsilon_m)\mathbf{E} - (\kappa^o_{\text{eff}} + \kappa^e_{\text{eff}})\mathbf{q} \times \mathbf{H}, \quad (5.61)$$

$$\mathbf{M} = \mu_0(\mu_{\text{eff}} - 1)\mathbf{H} + (\kappa^o_{\text{eff}} - \kappa^e_{\text{eff}})\mathbf{q} \times \mathbf{E}. \quad (5.62)$$

For a lattice of CPs this approach is also applicable if the "extraordinary" MEC parameter is a dyad performing the rotation by $\pi/2$ in the crystal plane $(x-y)$, i.e., κ^e is replaced by $\kappa^e_{\text{eff}}(\mathbf{x}_0\mathbf{y}_0 - \mathbf{y}_0\mathbf{x}_0)$ [25, 213].

Parameters ε_{eff}, μ_{eff} and κ^o_{eff} were expressed in works [24, 25] through the interaction factors of the lattice B_{me}, B_e, and B_m and individual polarizations. This analysis allowed the author to express his new EMPs through our old (first-principle) dynamic EMPs. In accordance with the modified homogenization model, the wave impedance Z_w and the refractive index n should be expressed through new ("effective") EMPs as follows:

$$Z_w = \sqrt{\frac{\mu_0\mu_{\text{eff}}}{\varepsilon_0\varepsilon_{\text{eff}}}}(\sqrt{1 - \xi^2_{\text{eff}}} + j\xi_{\text{eff}}), \quad (5.63)$$

where $\xi_{\text{eff}} = \omega\kappa^o_{\text{eff}}$ is the dimensionless MEC parameter of the Omega-type, and

$$n = \sqrt{\mu_{\text{eff}}\varepsilon_{\text{eff}}}\sqrt{1 - \xi^2_{\text{eff}}}. \quad (5.64)$$

Eqs. (5.63) and (5.64) are standard formulas for the wave impedance and refractive index of an Omega medium (see [79]). Using these equations and relations $Z_w = \sqrt{\mu_0\mu/\varepsilon_0\varepsilon}$ and $n = \sqrt{\varepsilon\mu}$ one may express "new" EMPs through our old ("equivalent") EMPs.

The linear q-dependence of the medium susceptibilities is definitely an effect of the phase shift per unit cell experienced by the eigenmode. This new type of WSD manifests as a seeming Omega-type bianisotropy of a non-bianisotropic MM. The Omega-type of MEC corresponds to the vector product in the second terms of the right-hand sides of Eqs. (5.59) and (5.60). Only this type of the fictitious bianisotropy can be attributed to the medium without a real bianisotropy. A fictitious chirality is not possible by an evident reason—chirality is an observable effect.

Let us present the main result in terms of the "new" and "old" EMPs using the notations adopted in the present book:

$$\varepsilon_{\text{new}} = \varepsilon_{\text{old}} \left(1 - \frac{\kappa_{\text{eff}}^o \omega}{n}\right), \quad \mu_{\text{new}} = \mu_{\text{old}} \left(1 - \frac{\kappa_{\text{eff}}^o \omega}{n}\right), \quad (5.65)$$

where $n \equiv q/k_0$ is the effective-medium refractive index that can be calculated using our dispersion equations or by full-wave simulations, and

$$\kappa_{\text{eff}}^o = \frac{q}{V} \frac{B_{em}}{(a_{ee}^{-1} - B_e)(a_{mm}^{-1} - B_m) - B_{em}^2}, \quad (5.66)$$

where different (more-or-less accurate) approximate formulas for interaction factors $B_e = B_m \eta^2$ and B_{em} can be found in works [15, 16, 18, 24, 97].

In Fig. 5.16(b) the frequency plots of these new EMPs are shown. Within the resonant stopband where the structure becomes a photonic crystal these parameters still do not make sense. However, at $k_0 a = [0.5-0.6]$ and $k_0 a = [0.65-0.7]$ the antiresonant behavior of our old permittivity is replaced by Lorentzian behavior of ε_{new}. The frequency dispersion of μ_{new} is also drastically different—a strange resonance of the old permeability at $k_0 a = 0.8$ is substituted by the resonant MEC parameter. The resonance of the MEC parameter is Lorentzian in the band $k_0 a = 0.65-0.7$.

The MEC parameter still turns out to be non-local (decreases versus frequency) in the band $k_0 a = 0.85-1.0$. This incorrect result is not a drawback of the new homogenization model. On the contrary, this wrong frequency behavior of κ_{eff} allows us to guess that a homogenization model is not applicable for this range. In fact, in this band the lattice is not effectively continuous and this frequency region should be also withdrawn from consideration as well as $k_0 a = 0.60$–0.65. The static effective permittivity of the medium formed by spheres of permittivity $\varepsilon_s = 120$ is equal $\varepsilon = 120 f + 1 - f$ where f is the fraction ratio of our spheres. For $D/a = 0.45$ the fraction ratio of the spheres is small— $f \approx 0.08$—however, their permittivity is huge. Therefore, the quasi-static effective-medium refractive index is much larger than unity— $n \approx 3.4$. This quasi-static estimate gives for the Bragg condition $2a = \lambda_{\text{eff}} = \lambda/2.4$ the normalized frequency $k_0 a = \pi/n = 0.9$. Around this frequency the MM cannot be homogenized.

Notice that at $k_0 a = 0.9$–0.95 the p-dipole Mie resonance occurs that destroys the Bragg resonance. In the part of the band $k_0 a = 0.85$–1.0, namely at $k_0 a = 0.9$–1.0 the eigenmode is propagating and $n > 0$. However, it does not mean that the lattice is effectively continuous, and the non-local MEC parameter obtained in the new model recalls us: When we calculate EMPs of a MM in a broad spectrum the prerequisite of the homogenization—optical density of the lattice—should be always checked. In the whole band $k_0 a = 0.85$–1.0 the homogenization is prohibited.

5.3.2.3 A lattice with low optical losses

In works [24, 213] the lattices were lossless. For example, in [213] the inclusions are magneto-dielectric spheres with $\varepsilon_s = 13.8$ and $\mu_s = 11$. Definitely, their "equivalent" EMPs also manifest the antiresonant behavior at the edges of the stopband. If we introduce very small losses for the material of spheres the results practically do not change up to the dielectric and magnetic loss tangent 0.001. However, if this tangent exceeds 0.01, the situation changes drastically. The author of this book has checked that for the structure studied in [24] the antiresonance of the "equivalent" permittivity disappears and the resonant stopband at $k_0 a = [0.6$–$0.65]$ replaces by the band of resonant losses. Both "equivalent" permittivity and permeability become Lorentzian functions of frequency in the whole considered frequency range. For such a lattice the fictitious bianisotropy is not relevant.

In work [25] the spheres with $\varepsilon_s = 13.8(1 - j0.007)$ and $\mu_s = 11$ manifest the antiresonance of the "equivalent" permittivity (can be seen in Fig. 3a of [25]) that is also cured by the fictitious bianisotropy. This is so because the tangent of dielectric loss 0.007 together with purely real and very high permeability corresponds to very low optical losses. Namely, the complex refractive index of the material of spheres is equal $n_s = \sqrt{\varepsilon_s \mu_s} \approx 12.3(1 - j0.002)$. The author of this book has checked that this antiresonance disappears in the case $n_s = 12.3(1 - j0.004)$.

In work [27] it was stressed that the fictitious bianisotropy of Omega type for symmetric lattices of non-bianisotropic particles is relevant for low-loss MMs and at two narrow frequency ranges located at the edges of the resonant stopband. At the center of the resonance band the resonant stopband of the Bragg type is located, the structure is a photonic crystal and cannot be homogenized in principle. The author of the present book stays on this point. In [27] it was experimentally shown that not only losses restore the physical meaning of first-principle EMPs. The impact of the low random aperiodicity of a p-m lattice is similar to that of losses. In slightly random MMs there is also no need to introduce the fictitious bianisotropy.

So, introduction of the effective bianisotropy for non-bianisotropic lattices and the corresponding modification of the permittivity and permeability may be useful only in the rather special case. This is the case of a low-loss and strictly regular p-m lattice operating at the frequencies located around the resonant stopband. This modification of the homogenization model is not a panacea. It does not allow us to homogenize an effectively discrete medium, e.g., the same lattice in its resonant stopband.

And it is worth to notice that even in the case when those new effective parameters are useful, their physical meaning is disputable. In fact, let us look at Fig. 2.3d depicting an OP. It is asymmetric, whereas the sphere emulated by this effective OP in the new model is symmetric. Formula (5.63) corresponds to the case when the wave propagates to the right, i.e., first impinges the arms of the OP. If the wave propagates to the left, it first impinges the loop. Therefore, formula (5.63) for a truly uniaxial Omega-medium gives the wave impedance depending on the propagation direction:

$$Z_w = \sqrt{\frac{\mu_0 \mu_{\text{eff}}}{\varepsilon_0 \varepsilon_{\text{eff}}}} (\sqrt{1 - \xi_{\text{eff}}^2} \pm j\xi_{\text{eff}}).$$

Sign plus corresponds to the propagation from the left to the right. Sign minus corresponds to the propagation from the right to the left. In any Omega-medium there is such a dependence of the wave impedance on the propagation direction expressing the asymmetry of the scatterer along the propagation path [79].

However, for a lattice of spheres described as an effectively uniaxial Omega-medium, there is no difference between these two cases, we have the same wave impedance for the waves propagating to the left and to the right. Thus, introducing the bianisotropy suggested in [24] we cannot identify it with any physically realizable medium. This is why we call this effective bianisotropy fictitious. These are arguments against this homogenization model. Should we apply this model for low-loss regular MMs of the Bloch type? Or is it better to decide that the first-principle dynamic homogenization is not adequate in the case of low losses. In the latter case we should ignore the bulk material parameters and, instead, characterize the medium via the Bloch impedance and refractive index. The author of the present book cannot advise one of two choices—it is a question of personal taste.

To conclude this chapter, we have to mention that an interesting experimental method allowing one to retrieve the first-principle dynamic EMPs ("equivalent" ones) was suggested in work [26]. This method is based on the theoretical or experimental analysis of the true electromagnetic fields inside the MM. For the experimental realization it implies the insertion of the small electric and magnetic probes into a MM under study. This insertion is possible without a significant perturbation of the propagating wave. The known results of this method are Lorentzian material parameters obtained in works [26, 214, 215] and coinciding with EMPs of our first-principle homogenization model. Notice that MMs studied in works [26, 214, 215] had sufficient losses preventing the Bragg stopband and the fictitious bianisotropy was not relevant for them.

Chapter 6

Conclusions

In this book a theory of weak spatial dispersion in bulk media composed of optically small scattering particles is presented in a comprehensive way. After a brief introductory chapter treating the basic concepts such as effective continuity and discontinuity, locality and non-locality, weak and strong spatial dispersion, etc., Chapter 2 presents the state of the art. This chapter reviews most relevant classical works in which quasi-static electromagnetic homogenization model suitable for media without WSD was developed. In this overview we discuss the difference between the phenomenological model adopted in the electrodynamics of continuous media, where the effects of weak spatial dispersion are introduced "by hands" and the homogenization model based on the local field approach where these effects result from the microscopic parameters—polarizabilities of the constitutive particles. The relevance of the homogenization theory where WSD results from the microscopic properties of the medium constituents is demonstrated on explicit examples of microwave composites. The microscopic content of both artificial magnetism and bianisotropy is revealed. Special attention is paid to the homogenization of the medium interface. The continuity of the interface is a key point for the solution of the boundary problems formulated for a homogenized medium

---

Composite Media with Weak Spatial Dispersion
Constantin Simovski
Copyright © 2018 Pan Stanford Publishing Pte. Ltd.
ISBN 978-981-4774-83-3 (Hardcover), 978-1-351-16624-9 (eBook)
www.panstanford.com

sample. For example, the continuity or discontinuity of the interface is important for the correct retrieval of the effective medium parameters from the scattering matrix of the medium layer.

A quasi-static theory based on the hierarchy of multipoles is developed in Chapter 3. It allows us to apply the homogenization model to finite-thickness layers of a composite or natural molecular media. In this model the classical Clausius–Mossotti–Lorentz–lorenz formulas relating the local and mean fields play an important role. They allow us to substitute the local field by the mean field in the medium polarization response. This theory aims the derivation of material equations of media with WSD and is finalized by a practical homogenization algorithm (generalized Maxwell Garnett mixing rules) relating the effective material parameters (entering these material equations) with the polarizabilities of the constitutive particles. The self-consistency of the whole theory including this homogenization procedure is confirmed by the Lorentzian and reciprocal behavior of the obtained EMPs.

Also, in Chapter 3 it is shown that there are three physical effects of WSD: artificial magnetism, bianisotropy and the second-order retardation effect described by the so-called b-parameter. The last effect corresponds to the presence of the spatial derivatives of the mean field in the material equations. All three effects are described in the book within the framework of the general multipolar model. With a numerical example the frequency bounds of WSD for non-resonant composites with artificial magnetism are outlined.

In Chapter 4 Maxwell's boundary conditions at the effective-medium interface are proved (and the location of this interface is specified). Here, the classical concept of transition layers initially introduced by P. P. Drude is revisited. It is shown that after an important correction the theory of transition layers by Drude becomes consistent with the experimental observations. The reason why the original Drude model contradicted the experiments is explained. The revised theory of the Drude transition layers not only corrects the mistake of the original model but also offers a significant improvement to the accuracy of the solution of the boundary problem for very thin composite slabs.

In the last chapter of this book (Chapter 5), we consider the homogenization of the media which possess the magnetic dipole resonance still being effectively continuous. Such composite media refer to the class of metamaterials. For MMs the rule of the spatial averaging for mean fields should be different from the classical quasi-static rule. For MMs with the magnetic resonance, the hierarchy of multipoles inherent to the non-resonant media with WSD is broken and Maxwell's boundary conditions are violated. It occurs at some frequencies of the resonance band or, sometimes, in the whole resonance band. Therefore, a new homogenization model is needed and this model is developed for the special case—regular MMs composed by electric (p) and magnetic (m) dipoles.

First, we homogenize an infinite p-m lattice. Our first-principle homogenization model results in two local EMPs: permittivity and permeability. Second, we separately consider the interface of a finite-thickness MM slab that is not effectively continuous anymore. Here, the idea of a transition layer is revisited once more. Now, it is not the Drude transition layer. The role of the new transition layer is to mimic the jump of the tangential components of the mean field vectors **E** and **H** at the physical interface of a MM. Effective parameters of such the transition layer are very different from those of the Drude layer.

EMPs of both bulk part of a MM slab and those of two transition layers located at its interfaces are retrieved in a self-consistent way resulting in the physical frequency behavior of all retrieved parameters. The retrieved permittivity and permeability of the bulk are exactly equal to those of the corresponding infinite lattice. This equivalence was achieved for the so-called Bloch lattices—orthorhombic lattices with weak near-field interaction between two crystal planes, adjacent along the propagation direction. To consider the interfaces of a resonant Bloch lattice as effectively boundaries becomes possible if we spread these boundaries into two transition layers. The bulk MM is sandwiched between them. Transition layers can be, in principle, compressed into infinitesimally thin effective sheets with electric and magnetic surface polarizations. However, this theory has not been sufficiently developed, yet.

Finally, an important group of MMs is analyzed, for which the first-principle homogenization theory must be revised. They are strongly resonant, low-loss, regular MMs of p-m dipoles. For these lattices the EMPs calculated in accordance with the first-principle model manifest the antiresonance. The reason for this erroneous result is analyzed. It is shown that for such MMs the antiresonance does not obviously mean the effective discontinuity. Although inside the resonant stop-band the MM is a photonic crystal, at the edges of the stop-band the MM is effectively continuous and can be homogenized so that its EMPs are physically sound. The revised homogenization model implies the redefined EMPs and introduces a fictitious bianisotropy of the effective medium, which allows us to get rid of the antiresonance in both frequency bands—below and above the resonance stop-band. We discuss the strong and weak points of this recently developed model.

The book is mainly dedicated to young researchers—PhD students and master students—specializing in metamaterials. Therefore, the most difficult part of the theory of regular MMs—the non-local homogenization and its implications—is intentionally omitted. Also, this book will be useful for a broad range of physicists and opticians eager to better understand the physical mechanism of weak spatial dispersion and its main manifestations: bianisotropy and artificial magnetism.

References

1. C. R. Simovski, *Weak Spatial Dispersion in Composite Media*, St. Petersburg, Polytechnica, 2003 (in Russian).
2. J. B. Pendry, Negative refraction makes a perfect lens, *Phys. Rev. Lett.* **85**, 3966–3969, 2000.
3. V. G. Veselago, The electrodynamics of substances with simultaneously negative values of ϵ and μ, *Sov. Phys. Uspekhi* **10**, 509–514, 1968.
4. F. Capolino, Editor, *Metamaterials Handbook, Vol. 1: Theory and Phenomena of Metamaterials*, CRC Press, London–NY, 2009. Chapter 2: Material parameters and field energy in reciprocal composite media.
5. L. D. Landau and E. M. Lifshitz, *Electrodynamics of Continuous Media*, Oxford University Press, Oxford, UK, 1982.
6. *Nanostructured Metamaterials*, A. de Baas, Editor-in-Chief, European Commission, Directorate General for Research, Publication Office of the European Union, B-1049 Brussels, 2010.
7. D. R. Smith and J. B. Pendry, Homogenization of metamaterials by field averaging, *JOSA B* **23**, 391–403, 2006.
8. D. R. Smith, S. Schultz, P. Markos and C. M. Soukoulis, Determination of effective permittivity and permeability of metamaterials from reflection and transmission coefficients, *Phys. Rev. B* **65**, 195104, 2002.
9. S. O'Brien and J. B. Pendry, Magnetic activity at infrared frequencies in structured metallic photonic crystals, *J. Phys.: Condens. Matter* **14**, 6383–6394, 2002.
10. C. R. Simovski, On the homogenization of artificial lattices, *Days on Diffraction'2006*, Steklov Institute, St. Petersburg, Russia, May 30–June 2, 2006, pp. 70–71.
11. C. Simovski, I. Kolmakov and S. Tretyakov, Approaches to the homogenization of periodical metamaterials, *Proc. MMET'06*, 11-th International Conference on Mathematical Methods in Electromagnetic Theory, Kharkov, Ukraine, June 26–29, 2006, pp. 41–44.

12. S. Tretyakov, C. Simovski and I. Kolmakov, Challenges in effective media modeling of artificial materials, *Third Workshop on Metamaterials and Special Materials for Electromagnetic Applications and TLC*, Rome, 30–31 March 2006, p. 26.
13. C. R. Simovski and S. A. Tretyakov, On effective material parameters of metamaterials, *Proc. 23-d Annual Review of Progress in Applied Computational Electromagnetics*, Verona, Italy, 19–23 March 2007, pp. 150–154.
14. C. R. Simovski and S. A. Tretyakov, Local constitutive parameters of metamaterials from an effective-medium perspective, *Phys. Rev. B* **75**, 195111(1–9), 2007.
15. C. R. Simovski, Bloch material parameters of magneto-dielectric metamaterials and the concept of Bloch lattices, *Metamaterials* **1**, 62–80, 2007.
16. C. R. Simovski, Analytical modelling of double-negative composites, *Metamaterials* **2**, 169–185, 2008.
17. C. R. Simovski, On material parameters of metamaterials (review), *Optics and Spectroscopy* **107**, 726–753, 2009.
18. M. G. Silveirinha, Metamaterial homogenization approach with application to the characterization of microstructured composites with negative parameters, *Phys. Rev. B* **75**, 115104, 2007.
19. M. G. Silveirinha, Poynting vector, heating rate, and stored energy in structured materials: A first-principles derivation, *Phys. Rev. B* **80**, 235120, 2009.
20. S. B. Glybovski, S. A. Tretyakov, P. A. Belov, Y. S. Kivshar and C. R. Simovski, Metasurfaces: From microwaves to visible, *Phys. Rep.* **634**, 1–72, 2016.
21. S. Kim, E. F. Kuester, C. L. Holloway, A. D. Sher and J. Baker-Jarvis, Boundary effects on the determination of metamaterial parameters from normal incidence reflection and transmission measurements, *IEEE Trans. Antennas Propag.* **59**, 2226–2240, 2011.
22. S. Kim, E. F. Kuester, C. L. Holloway, A. D. Sher and J. Baker-Jarvis, Effective material property extraction of a metamaterial by taking boundary effects into account at TE/TM polarized incidence, *Prog. Electromagn. Res. B* **36**, 1–33, 2012.
23. A. P. Vinogradov, A. I. Ignatov, A. M. Merzlikin, S. A. Tretyakov and C. R. Simovski, Additional effective-medium parameters for composite materials – excess surface currents, *Opt. Express* **19**, 6699–6704, 2011.

24. A. Alù, Restoring the physical meaning of metamaterial constitutive parameters, *Phys. Rev. B* **83**, 081102, 2011.
25. A. Alù, First-principles homogenization theory for periodic metamaterials, *Phys. Rev. B* **84**, 075153, 2011.
26. A. Andryieuski, S. Ha, A. A. Sukhorukov, Y. S. Kivshar and A. V. Lavrinenko, Unified approach for retrieval of effective parameters of metamaterials, *Proc. SPIE* **8070**, 807008, 2011.
27. P. Alitalo, A. E. Culhaoglu, C. R. Simovski and S. A. Tretyakov, Experimental study of anti-resonant behavior of material parameters in periodic and aperiodic composite materials, *J. Appl. Phys.* **113**, 224903, 2013.
28. S. L. Donaldson and D. B. Miracle, Editors, *Engineered Materials Handbook, Volume 21: Composites*, ASM International, Delaware City, DE, 1987.
29. M. Taya and J. R. Vinson, Editors, *Recent Advances in Composites in USA and Japan*, Issue No. 864, ASTM International Publishers, NY, 1985.
30. C. Zweben, *Advanced Materials for Optoelectronics and MEMs Package Set: Parts 1 and 2*, SPIE Press, Bellingham, WA, 2002.
31. G. Ruck, Editor, *Radar Cross Section Handbook*, Plenum, NY, 1970.
32. R. Chatterjee, *Antenna Theory and Practice*, 2nd ed., New Age International Ltd Publishers, New Delhi, 1996.
33. C. Checkroun, RADANT: New method of electronic scanning, *Microw. J.* **17**, 78–85, 1981.
34. J. J. Mäki, M. Kauranen and A. Persoons, Surface second-harmonic generation from chiral materials, *Phys. Rev. B* **51**, 1425–1433, 1995.
35. J. D. Joannopoulos, R. D. Mead and J. N. Winn, *Photonic Crystals*, Princeton University Publishing, Princeton, USA, 1995.
36. Y. A. Vlasov, M. Deutsch, D. J. Norris, Focusing on single photonic crystal, *Appl. Phys. Lett.* **76**, 1627–1629, 2000.
37. K. Busch, S. Lölkes, R. B. Wehersponhm and H. Föll, Editors, *Photonic Crystals: Advances in Design, Fabrication and Characterization*, Wiley-VCH, Berlin, 2004.
38. L. D. Barron, *Molecular Light Scattering and Optical Activity*, University Press, Cambidge, UK, 1982.
39. E. B. Graham and R. E. Raab, Molecular scattering in spatially dispersive medium, *Proc. R. Soc. London A*, **430**, 593–614, 1990.
40. F. I. Fedorov, *Theory of Gyrotropy*, Nauka i Tekhnika, Minsk, USSR, 1976 (in Russian).

41. E. J. Post, *Formal Structure of Electromagnetics*, North-Holland Publishing Company, Amsterdam, 1962.
42. S. R. de Groot and L. G. Suttorp, *Foundations of Electrodynamics*, North-Holland Publishing Company, Amsterdam, 1972.
43. A. Lakhtakia, B. Michel, W. S. Weighlhofer. Bruggeman model for bianisotropic columnar films, *Optics Comm.* **65**, 3804–3814, 1998.
44. B. E. Perilloux, *Thin Film Design*, SPIE Press, Bellingham, WA, 2002.
45. B. V. Bokut, A. N. Serdyukov and F. I. Fedorov, The phenomenological theory of optically active crystals. *Soviet Phys. Crystallogr.* **15**, 871–874 (1971).
46. E. F. Knott, J. F. Shaeffer and M. T. Tuley, *Radar Cross Section*, 2nd ed., SciTech Publishing, Raleigh, NC, 2004.
47. S. A. Schelkunoff and H. T. Friis, *Antenna Theory and Practice*, Wiley Interscience, New York, 1952.
48. A. Raffo and G. Crupi, *Microwave Wireless Communications: From Transistor to System Level*, Elsevier Academic Press, Amsterdam, 2016.
49. O. Dernovsek, M. Eberstein, W.A Schiller, A. Naeini, G. Preu, W. Wersing, LTCC glass-ceramic composites for microwave application, *J. Eur. Ceramic Soc.* **21**, 1693–1697, 2001.
50. A. N. Lagarkov and K. N. Rozanov, High-frequency behavior of magnetic composites, *J. Magn. Magn. Mater.* **321**, 2082–2092, 2009.
51. S. I. Pekar, *Optics of Crystals and Additional Light Waves*, Kiev, Naukova Dumka, 1982 (in Russian).
52. A. Sommerfeld, *Optics*, Academic Press, NY, 1954.
53. J. D. Jackson, *Classical Electrodynamics*, 3rd ed., Wiley Interscience, NY, 1998.
54. J. Schwinger, K. A. Milton, L. L. DeRaad and W.-Y. Tsai, *Classical Electrodynamics*, 3rd ed., Perseus, NY, 1998.
55. A. R. von Hippel, *Dielectrics and Waves*, Artech House, London, 1995.
56. R. E. Raab and O. L. de Lange, *Multipole Theory in Electromagnetism*, Clarendon Press, London, 2005.
57. D. Felbacq and G. Bouchitté, Theory of mesoscopic magnetism in photonic crystals, *Phys. Rev. Lett.* **94**, 2005, 183902.
58. A. I. Cabuz, D. Felbacq and D. Cassagne Homogenization of negative-index composite metamaterials: A two-step approach, *Phys. Rev. Lett.* **98**, 037403, 2007.

59. K. Vynck, D. Felbacq, E. Centeno, A. I. Cabuz, D Cassagne and B Guizal, All-dielectric rod-type metamaterials at optical frequencies, *Phys. Rev. Lett.* **102**, 133901, 2009.
60. I. Tsukerman, Effective parameters of metamaterials: A rigorous homogenization theory via Whitney interpolation, *J. Opt. Soc. Am. B* **28**, 577–585, 2011.
61. Y. Chen and R. Lipton, Resonance and double negative behavior in metamaterials, *Arch. Rational Mech. Anal.* **209**, 835–868, 2013.
62. R. A. Shore and A. D. Yaghjian, Complex waves on periodic arrays of lossy and lossless permeable spheres: Part 1. Theory, *Radio Sci.* **47**, RS2014, 2012.
63. A. P. Vinogradov, *Electrodynamics of Composite Media*, URSS Publishing, Moscow, 2001 (in Russian).
64. M. Born and E. Wolf, *Principles of Optics*, Pergamon Press, London, 1959.
65. M. Born and H. Kun, *Dynamic Theory of Crystal Lattices*, Oxford University Press, Oxford, UK, 1954.
66. R. W. Munn, Microscopic theory of dielectric response for molecular multilayers, *J. Chem. Phys.* **99**, 1404–1408, 1993.
67. P. P. Drude, *The Theory of Optics*, Logmans, Green and Co Publishers, NY, 1902.
68. H. A. Kramers, La diffusion de la lumiére par les atomes, in: *Atti Congressi Internazionale Fisici da Luigi Volta (Transactions of Volta Centenary Congress in Physics)*, Como, Italy, May 1927, Vol. 2, pp. 545–557.
69. R. de Ludwig Kronig, On the theory of the dispersion of X-rays, *J. Opt. Soc. Am.* **12**, 547–557, 1926.
70. P. Shekhar, A. Atkinson and Z. Jacob, Hyperbolic metamaterials: fundamentals and applications, *Nano Convergence* **1**, 14, 2014.
71. N. A. Krall and A. W. Trivelpiece, *Principles of Plasma Physics*, McGraw Hill, NY, 1973.
72. D. R. Smith, W. J. Padilla, D. C. Vier, S. C. Nemat-Nasser and S. Schultz, Composite media with simultaneously negative permeability and permittivity, *Phys. Rev. Lett.* **84**, 4184–4187, 2000.
73. R. A. Shelby, D. R. Smith and S. Schultz, Experimental verification of a negative index of refraction, *Science* **292**, 77–79, 2001.

74. H. A. Lorentz, *The Theory of Electrons and Its Applications to the Phenomena of Light and Radiant Heat*, B. G. Teubner Publisher, Leipzig, Germany, 1916.
75. V. M. Agranovich and V. L. Ginzburg, *Spatial Dispersion in Crystal Optics and the Theory of Excitons*, Wiley InterScience, NY, 1966.
76. W. A. Davis and C. M. Krowne, The effect of drift and diffusion in semiconductors on plane wave interaction at interfaces. *IEEE Trans. Antennas Propag.* **36**, 97–103, 1988.
77. A. H. Sihvola and I. V. lindell, Effective permeability of mixtures, *Prog. Electromagn. Res.* **6**, 412–423, 1994.
78. B. D. H. Tellegen, The gyrator: A new electric network element, *Philips Res. Rep.* **3**, 81–94, 1948.
79. A. N. Serdyukov, I. V. Semchenko, S. A. Tretyakov and A. Sihvola, *Electromagnetics of Bianisotropic Materials: Theory and Applications*, Gordon and Breach, Amsterdam, 2001.
80. I. V. Lindell, A. H. Sihvola, S. A. Tretyakov and A. J. Viitanen, *Electromagnetic Waves in Chiral and Bi-Isotropic Media*, Artech House, Boston–London, 1994.
81. L. Onsager. Reciprocal relations in irreversible processes, Part I, *Phys. Rev.* **37**, 405–426, 1931. Part II, *ibid.* **38**, 2265–2279, 1931.
82. E. O. Kamenetskii, Magnetostatically controlled bianisotropic media: A novel class of artificial magnetoelectric materials, in: *Advances in Complex Electromagnetic Materials*, A. Priou, A. Sihvola, S. Tretyakov and A. Vinogradov, Editors. Kluwer Academy Press, Dordrecht, 1997, pp. 412–423.
83. F. Guerin, Microwave chiral materials: A review of experimental studies and some results on composites with ferroelectric ceramic inclusions, *Prog. Electromagn. Res.* **9**, 219–263, 1994.
84. A. Lakhtakia, B. Michel, W. S. Weighlhofer. Bruggeman formalism for two models of uniaxial composite medium, *Composites Sci. Technol.* **57**, 185–196, 1997.
85. E. B. Graham and R. E. Raab, Molecular multipole moments and macroscopic electric field, *Proc. R. Soc. London A*, **430**, 593–614, 1990.
86. R. E. Raab, Multipole moments of long molecules and multipolar polarization of molecular media, *Mol. Phys.* **29**, 1323–1331, 1975.
87. E. B. Graham and J. Pierrus, Multipole moments and Maxwell equations, *J. Phys. B* **25**, 4673–4684, 1992.

88. A. N. Godlevskaya, Some peculiarities of spehrical electromagnetic waves propagation in media with natural gyrotropy, *Reports of the Belarusian Academy of Sciences* **33**, 616–618, 1987.
89. A. Lakhtakia and W. S. Weighlhofer, Are field derivatives needed in linear constitutive equations? *Int. J. Infrar. Millim. Waves* **19**, 1073–1083, 1998.
90. A. A. Maradudin and D. L. Mills, Effect of spatial dispersion on the properties of a semi-infinite dielectric, *Phys. Rev. B* **7**, 2787–2794, 1973.
91. C. R. Simovski, First-order spatial dispersion and Maxwell Garnett modeling of bianisotropic media, *Int. J. Electron. Commun.* **52**, 76–81, 1998.
92. S. I. Maslovski, C. R. Simovski and S. A. Tretyakov, Constitutive equations for media with second-order spatial dispersion, *Proceedings of the 6th International Workshop on Chiral, Biisotropic and Bianisotropic Media Bianisotropics'98*, Braunschweig, May 28–31, 1998, pp. 36–39.
93. C. R. Simovski, On the averaging of induced multipoles in media with weak spatial dispersion, *Int. J. Electron. Commun.* **54**, 271–274, 2000.
94. C. R. Simovski, S. He and M. Popov, On the dielectric properties of thin molecular or composite layers, *Phys. Rev. B* **62**, 13168–13176, 2000.
95. C. R. Simovski, S. A. Tretyakov, A. H. Sihvola and M. Popov, On the surface effects in thin molecular layers, *Eur. Phys. J.: Appl. Phys.* **9**, 233–240, 2000.
96. C. R. Simovski and B. Sauviac. On the bulk averaging approach for obtaining the effective parameters of thin magnetic granular films, *Eur. Phys. J.: Appl. Phys.* **17**, 11–20, 2002.
97. C. R. Simovski and S. He, Frequency range and explicit expressions for negative permittivity and permeability of an isotropic medium formed by a lattice of perfectly conducting Omega-particles, *Phys. Lett. A* **311**, 254–263, 2003.
98. C. R. Simovski, Application of the Fresnel formulas for reflection and transmission of electromagnetic waves beyond the quasi-static approximation, *J. Commun. Tech. Electron.* **52** 953–967, 2007.
99. A. P. Vinogradov, On formulas by Clausius, Mossotti, Lorentz and Lorenz for magnetic media, *J. Commun. Technol. Electron.* **44**, 1131–1139, 1999.
100. L. V. Lorenz, Sur la lumiére réfléchie et réfractée par une sphére transparente, in: *Oeuvres scientifiques de L. Lorenz*, Volume 1, pp. 405–529, Fb. and C. Limited, Copenhagen, 1898.

101. E. Madelung, Elektromagnetische Theorie für Reflexion aus den Kristallen, *Physik ZS* **19**, 524–536, 1912.
102. P. P. Ewald, Zur Begründung der Kristalloptik: Teil 3, *Ann. Phys.* **54**, 519–597, 1917.
103. P. P. Ewald, Zur Begründung der Kristalloptik: Teil 1, *Ann. Phys.* **49**, 1–38, 1916, Teil 2, *Ann. Phys.* **49**, 117–143, 1916.
104. C. W. Oseen, Optische Strahlugsfelders in der Kristallen, *Annalen der Physik* **48**, 1–21, 1915.
105. Lord Rayleigh, On the dynamic theory of gratings, *Proc. R. Soc. London* **A79**, 399–416, 1907.
106. G. Russakoff, A derivation of the macroscopic Maxwell equations, *Am. J. Phys.* **38**, 1188–1195, 1970.
107. J. E. Sipe, J. Van Kranendonk, On the light reflection from polarised layered structures, *Phys. Rev. A.* **53**, 1806–1822, 1974.
108. M. R. Phillpott, On the light-induced polarisation of the transition surface layer, *J. Chem. Phys.* **60**, 1410–1419, 1974.
109. J. E. Stewart and W. S. Gallaway. Diffraction anomalies in gratings spectrometry, *Appl. Opt.* **1**, 421–429, 1962.
110. A. P. Kazantsev, G. I. Surdutovich and V. P. Yakovlev. *Mechanical Action of Light on Atoms*. J. Wiley and Sons, NY, 1990.
111. M. R. Phillpott, Transverse electric field in the theory of surface polaritons, *J. Chem. Phys.* **60**, 2520–2529, 1974.
112. T. G. Mackay and W. S. Weiglhofer, Homogenization of biaxial composite materials: Dissipative anisotropic properties, *J. Opt. A: Pure Appl. Opt.* **2**, 426–432 (2000).
113. G. D. Mahan and H. Obermair, Polaritons at surfaces, *Phys. Rev.* **183**, 834–841, 1969.
114. P. A. Belov and C. R. Simovski, Boundary conditions for interfaces of electromagnetic (photonic) crystals and generalized Ewald-Oseen extinction principle, *Phys. Rev. B* **73**, 045102, 2006.
115. K.-N. Liou and P. Yang, *Light Scattering by Ice Crystals*, Cambridge University Press, Campbridge, UK, 2016.
116. D. V. Sivukhin, Molecular theory of reflection and refraction of light in liquids, *Zhurnal Experimentalnoi i Tekhnicheskoi Fiziki* **18**, 976–994, 1948 (in Russian).
117. D. V. Sivukhin, Elliptic polarization of light reflected from liquids, *Sov. Phys. JETP* **3**, 269–276, 1956.

118. P. P. Drude, Über die bei der Reflexion an isotropen, durchsichtigen Medien, *Wied. Ann.* **43**, 146, 1891.

119. C. V. Raman and L. A. Ramdas, The scattering of light by liquid boundaries and its relation to surface tension, *Proc. R. Soc. (London), Series A* **108**, 561–567, 1925.

120. C. V. Raman and L. A. Ramdas, On the thickness of the optical transition layer in liquid surfaces, *Philos. Mag.* **3**, 220–224, 1927.

121. O. S. Heavens, *Optical Properties of Thin Solid Films*, Butterworths Scientific Publications, London, 1965.

122. R. Landauer, Electric transport and optical properties of inhomogeneous media, *AIP Conferences Proc.*, Issue 40, NY, Oct. 1978, pp. 2–20.

123. R. E. Collin, *Field Theory of Guided Waves*, 3rd ed., Wiley and Sons, NY, 2001, Chapter 12.

124. A. Yagjian, Electric dyadic Green functions in the source region, *Proc. IEEE* **68**, 248–263, 1980.

125. R. W. P. King and G. S. Smith, *Antennas in Matter: Fundamentals, Theory, and Applications*, MIT Press, Cambridge, MA, 1981, Section 6.6.

126. Y. Wu, J. Li, Z.-Q. Zhang and C. T. Chan, Effective medium theory for magnetodielectric composites: Beyond the long-wavelength limit, *Phys. Rev. B* **74**, 085111, 2006.

127. X. Zhang and Y. Wu, Effective medium theory for anisotropic metamaterials, *Sci. Rep.* **5**, 7892, 2015.

128. R. Evans, Maxwell's color photography, *Sci. Am.* **205**, 117–128, 1961.

129. J. C. Maxwell Garnett, Colours in metal glasses and metallic films, *Phil. Trans. R. Soc. (London), Series A* **203**, 385–420, 1904.

130. J. C. Maxwell Garnett, Colours in metal glasses, in metallic films, and in metallic solutions, *Phil. Trans. R. Soc. (London), Series A* **205**, 237–288, 1906.

131. J. B. Pendry, A. J. Holden, D. J. Robbins and W. J. Stewart, Magnetism from conductors and enhanced nonlinear phenomena, *IEEE Trans. Microw. Theory Tech.* **47**, 2075–2083, 1999.

132. A. A. Sochava, C. R. Simovski and S. A. Tretyakov, Chiral effects and eigenwaves in bi-anisotropic omega structures, in: *Advances in Complex Electromagnetic Materials*, A. Priou, A. Sihvola, S. Tretyakov and A. Vinogradov, Editors, Kluwer Academic Publishers, Dordrecht-Boston-London, pp. 85–102, 1997.

133. S. A. Tretyakov, C. R. Simovski and A. A. Sochava, The relation between co-and crosspolarizabilities of small conductive bi-anisotropic

particles, in: *Advances in Complex Electromagnetic Materials*, A. Priou, A. Sihvola, S. Tretyakov and A. Vinogradov, Editors, Kluwer Academic Publishers, Dordrecht-Boston-London, pp. 271–280, 1997.

134. K. F. Lindman. Zur electrischer Leitfahigkeit metallischer Aggregate, *Ann. Phys.* **63**, 621–626, 1920.

135. N. Engheta and D. L. Jaggard, Waveguides using chiral materials, Patent US 50165059 A, 1992.

136. M. M. I. Saadoun and N. Engheta, A reciprocal phase shifter, using novel pseudochiral or omega medium, *Microw. Optics Tech. Lett.* **5**, 184–186, 1992.

137. S. A. Tretyakov and F. Mariotte, Maxwell Garnett modeling of uniaxial chiral composite media, *J. Electromagn. Waves Appl.* **9**, 1073–1089, 1995.

138. C. Monzon, New reciprocity theorems for chiral, nonactive and biisotropic media, *IEEE Trans. Microw. Theory Tech.*, **44**, 2299–2301, 1996.

139. F. Mariotte, S. A. Tretyakov and B. Sauviac. Isotropic chiral composite modeling: comparison with analytical, numerical and experimental data, *Microw. Optics Tech. Lett.* **7**, 861–864, 1994.

140. T. Guire, V. V. Varadan and V. K. Varadan. Experimental study of chiral composite medium for microwave applications, *IEEE Trans. Electromagn. Compat.* **EMC-32**, 300–309, 1990.

141. T. G. Kharina, S. A. Tretyakov, A. A. Sochava, C. R. Simovski and S. Bolioli, Experimental study of artificial omega media, *Electromagnetics* **20** 428–442, 1998.

142. I. P. Theron and J. H. Cloete. The optical activity of an artificial uniaxial chiral crystals at microwaves, *J. Electromagn. Waves Appl.* **10**, 539–562, 1996.

143. D. A. G. Bruggeman, Berechnung verschiedener physikalischer Konstanten von heterogenen Substanzen, I: Dielektrizitätskonstanten und Leitfähigkeiten der mischkörper aus isotropen Substanzen, *Annalen der Physik* **416**, 636–664, 1935.

144. A. Lakhtakia, *Selected Papers in Linear Optical Composite Materials*, SPIE Opt. Tech. Press, Bellingham, WA, 1996.

145. *Langmuir-Blodgett Films*, G. G. Roberts, Editor, Plenum Press, NY, 1990.

146. X. Lafosse, Preparation of new chiral composite with conductive polymers and its free space characterization, in: *Proc. Prog. Electromag. Res. Symposium PIERS'94*, Noordwijk, NL, pp. 53–56, 1994.

147. A. G. Fokin, Macroscopic conductivity of randomly hetero-homogeneous media: Methods of calculation, *Physics Uspekhi* **166**, 1070–1093, 1996.
148. D. E. Aspnes, Local-field effects and effective-medium theory: A macroscopic perspective, *Am. J. Phys.* **50**, 704–709, 1982.
149. A. N. Lagarkov, A. K. Sarychev, I. R. Smychkovich, A. P. Vinogradov, Effective medium theory for microwave dielectric constant and magnetic permeability of conducting stick composites. *J. Electromagn. Waves Appl.* **6** 1159–1176, 1992.
150. C. R. Simovski, M. S. Kondratjev, P. A. Belov and S. A. Tretyakov, Interaction effects in two-dimensional bianisotropic arrays, *IEEE Trans. Antennas Propag.* **47**, 1429–1439 (1999).
151. C. R. Simovski, P. A. Belov, M. S. Kondratiev, Electromagnetic interaction in regular arrays of uniaxial chiral scatterers. *J. Electromagn. Waves Appl.* **13**, 189–204 (1999).
152. L. Lewin, The electrical constants of a structure comprising two different dielectric sphere lattices embedded in a dielectric matrix *Proc. Inst. Elec. Eng.* **94**, 65–81, 1947.
153. C. A. Grimes and D. M. Grimes. The effective permeability of granular thin films, *IEEE Trans. Magn.* **29**, 4092–4094, 1993.
154. L. Jylhä and A. Sihvola, Equation for the effective permittivity of particle-filled composites for material design applications, *J. Phys. D: Appl. Phys.* **40**, 4966–4973, 2007.
155. V. Yannopapas and N. V. Vitanov, Photoexcitation-induced magnetism in arrays of semiconductor nanoparticles with a strong excitonic oscillator strength, *Phys. Rev. B* **74**, 193304, 2006.
156. M. S. Wheeler, J. S. Aitchison and M. Mojahedi, Three-dimensional array of dielectric spheres with an isotropic negative permeability at infrared frequencies, *Phys. Rev. B* **72**, 193103, 2005.
157. Q. Zhao, J. Zhou, F. Zhang and D. Lippens, Mie resonance-based dielectric metamaterials, *Mater. Today* **12**, 60–69, 2009.
158. C. L. Holloway, E. F. Kuester, J. Baker-Jarvis and P. Kabos, A double negative composite medium composed of magnetodielectric spherical particles embedded in a matrix, *IEEE Trans. Antennas Propag.* **51**, 2596–2603, 2003.
159. L. Peng, L. Ran, H. Chen, H. Zhang, J. A. Kong and T. M. Grzegorczyk, Experimental observation of left-handed behavior in an array of standard dielectric resonators, *Phys. Rev. Lett.* **98**, 157403, 2010.

160. A. V. Ghiner and G. I. Surdutovich, Discreteness and local fields in weakly rarefied media, *Phys. Rev. E* **56**, 6123–6130, 1997.
161. A. D. Buckingham and P. J. Stiles, On the multipolar re-radiation of light by molecules, *Mol. Phys.* **24**, 99–108, 1972.
162. S. A. Tretyakov, I. S. Nefedov, C. R. Simovski and S. I. Maslovski, Modelling and microwave properties of artificial materials with negative parameters, in: *Advances in Electromagnetics of Complex Media and Metamaterials*, S. Zouhdi, A. Sihvola and M. Arsalane, Editors, Kluwer Academy Publishers, Dordrecht, 2002, pp. 99–122.
163. R. E. Raab and J. H. Cloete, Circular birefriengence and dichroism in non-magnetic chiral media, *J. Electromagn. Waves Appl.* **8**, 1073–1089, 1994.
164. E. B. Graham and R. E. Raab, Covariant multipole D and H fields for reflection from a magnetic anisotropic chiral medium, in: *Advances in Complex Electromagnetic Materials*, A. Priou, A. Vinogradov and S. Tretyakov, Editors, Kluwer Academy Publishers, Dordrecht-Boston-London, 1997, pp. 55–67.
165. A. D. Buckingham and M. B. Dunn, A relationship for some multipolar susceptibilities of molecules. *J. Chem. Soc. A* **14**, 1988–1991, 1971.
166. H. Chen, L. Ran, J. Huangfu, X. M. Zhang, K. Chen, T. M. Grzegorczyk, and J. A. Kong, Left-handed materials composed of only S-shaped resonators, *Phys. Rev. E* **70**, 057605, 2004.
167. K. Aydin, I. Bulu, K. Guven, M. Kafesaki, C. M. Soukoulis and E. Ozbay, Investigation of magnetic resonances for different split-ring resonator parameters and designs, *N. J. Phys.* **7**, 168–180, 2005.
168. A. Ishikawa, T. Tanaka and S. Kawata, Negative magnetic permeability in the visible light region, *Phys. Rev. Lett.* **95**, 237401, 2005.
169. D. E. Logan, On the reciprocity and duality principles in molecular scattering theory, *Mol. Phys.* **46**, 271–285, 1982.
170. B. Sauviac, C. R. Simovski and S. A. Tretyakov, Double split-ring resonators: Analytical modeling and numerical simulations, *Electromagnetics* **24**, 317–338, 2004.
171. A. N. Grigorenko, A. K. Geim, H. F. Gleeson, Y. Zhang, A. A. Firsov, I. Y. Khrushchev and J. Petrovic, Nanofabricated media with negative permeability at visible frequencies, *Nature* **438**, 335–339, 2005.
172. G. Dolling, C. Enkrich, M. Wegener, S. Linden J. Zhou and C. M. Soukoulis, Cut-wire and plate capacitors as magnetic atoms for optical metamaterials, *Opt. Lett.* **30**, 3198–3200, 2005.

173. J. Zhou, L. Zhang, G. Tuttle, T. Koschny and C. M. Soukoulis, Negative index materials using simple short wire pairs, *Phys. Rev. B* **73**, 041101, 2006.
174. C. R. Simovski and S. A. Tretyakov, On effective electromagnetic parameters of artificial nano-structured magnetic materials, *Photon. Nanostruct. Fundam. Appl.* **8**, 254–263, 2010.
175. S. Linden, C. Enkrich, M. Wegener, J. F. Zhou, T. Koschny and C. M. Soukoulis, Magnetic response of metamaterials at 100 terahertz, *Science* **306** 1351–1354, 2004.
176. N. Katsarakis, G. Konstantinidis, A. Kostopoulos, R. S. Penciu, T. F. Gundogdu, T. Koschny, M. Kafesaki, E. N. Economou and C. M. Soukoulis, Magnetic response of split-ring resonators in the far infrared frequency regime, *Opt. Lett.* **30**, 1348–1350, 2005.
177. C. Enkrich, S. Linden, M. Wegener, S. Burger, L. Zswchiedrich, F. Schmidt, J. Zhou, T. Koschny and C. M. Soukoulis, Magnetic metamaterials at telecommunication and visible frequencies, *Phys. Rev. Lett.* **95**, 203901, 2005.
178. S. Zhang, W. Fan, N. C. Panoiu, K. M. Malloy, R. M. Osgood and S. R. J. Brueck, Experimental demonstration of near-infrared negative-index metamaterials, *Phys. Rev. Lett.* **95**, 137404, 2005.
179. J. Zhou, T. Koschny, M. Kafesaki, E. N. Economou, J. B. Pendry and C. M. Soukoulis, Limit of the negative magnetic response of split-ring resonators at optical frequencies, *Phys. Rev. Lett.* **95**, 223902, 2005.
180. K. C. Huang, M. L. Povinelli and J. D. Joannopoulos, Negative effective permeability in polaritonic photonic crystals, *Appl. Phys. Lett.* **85**, 543–545, 2004.
181. S. Linden, M. Decker and M. Wegener, Model system for a one-dimensional magnetic photonic crystal, *Phys. Rev. Lett.* **97**, 083902, 2006.
182. V. M. Agranovich, Y. N. Gartstein, Spatial dispersion and negative refraction of light, *Phys. Usp.* **49**, 10–44, 2006. 10.
183. M. G. Silveirinha, Metamaterial homogenization approach with application to the characterization of microstructured composites with negative parameters, *Phys. Rev. B* **75**, 115104, 2007.
184. Lord Rayleigh, On the influence of obstacles arranged in rectangular order upon the properties of a medium. *Philos. Mag.* **34**, 481–502, 1892.
185. A. A. Sochava, C. R. Simovski and S. A. Tretyakov, Chiral effects and eigenwaves in bi-anisotropic Omega structures, in: *Advances in*

Complex Electromagnetic Materials, A. Priou, A. Vinogradov and S. Tretyakov, Editors, Kluwer Academy Publishers, Dordrecht-Boston-London, 1997, pp. 85–103.

186. Y. M. and V. Meriakri, High-permittivity microwave dielectrics, *Electromagnetic waves and electronic systems* **2**, 35–44, 1997.

187. S. F. Karmanenko, A. D. Kanareikyn, E. A. Nenasheva, A. I. Dedyk and A. A. Semenov, Frequency dependence of microwave quality factor of doped (Ba,Sr)TiO$_3$ ferroelectric ceramics, *Integr. Ferroelectr.* **61**, 177–181, 2004.

188. Y. Xu and B. A. S. Gustafson, A generalized multiparticle Mie-solution: further experimental verification, *J. Quant. Spectrosc. Radiative Transfer* **70**, 395–419, 2001.

189. S. Enoch and N. Bonod, Editors, *Plasmonics: From Basics to Advanced Topics*, Springer, 2012. Chapter 2: Theory of wood anomalies.

190. S. A. Tretyakov and A. J. Viitanen, Plane waves in regular arrays of dipole scatterers and effective medium modelling, *J. Opt. Soc. Am. A* **17**, 1791–1799, 2000.

191. C. R. Simovski, P. A. Belov, A. V. Atraschenko and Y. S. Kivshar, Wire Metamaterials: Physics and Applications, *Adv. Mater.* **24**, 4229–4248, 2012.

192. L. Jylhä, I. Kolmakov, S. Maslovski and S. Tretyakov, Modeling of isotropic backward-wave materials composed of resonant spheres, *J. Appl. Phys.*, **99**, 043102, 2006.

193. J. B. Pendry and D. R. Smith, Reversing light with negative refraction, *Phys. Today* **57**, 37–38, 2004.

194. F. Capolino, Editor, *Metamaterials Handbook, Vol. 2: Applications of Metamaterials*, CRC Press, London–NY, 2009. Chapter 2: Flat lenses formed by photonic and electromagnetic crystals.

195. C. R. Simovski and B. Sauviac, Towards creating isotropic microwave composites with negative refraction, *Radio Sci.* **39**, 2014–2032, 2004.

196. J. D. Baena, L. Jelinek and R. Marques, Towards a systematic design of isotropic bulk magnetic metamaterials using the cubic point groups of symmetry, *Phys. Rev. B* **76**, 245115, 2007.

197. D. Morits and C. R. Simovski, Isotropic negative refractive index at near infrared, *J. Opt.* **14**, 125102, 2012.

198. A. M. Nicholson and G. F. Ross, The measurement of the intrinsic properties of materials by time-domain techniques, *IEEE Trans. Instrum. Meas.* **17**, 395–401, 1968.

199. W. W. Weir, Automatic measurement of the complex dielectric constant and permeability at microwave frequencies, *Proc. IEEE*, **62**, 33–40, 1974.
200. L. P. Ligthart, A fast computational technique for accurate permittivity and permeability retrieval using transmission line methods, *IEEE Trans. Microw. Theory Tech.* **31**, 249–257, 1983.
201. J. Baker-Jarvis, E. J. Vanzura and W. A. Kissick, Improved technique for determining complex permittivity with the transmission/reflection method, *IEEE Trans. Microw. Theory Tech.* **38**, 10961–10965, 1990.
202. X. Chen, T. Grzegorczyk, B.-E. Wu, J. Pacheko and J. A. Kong, Robust method to retrieve the constitutive effective parameters of metamaterials, *Phys. Rev. E* **70**, 016608, 2004.
203. N. Katsarakis, T. Koschny, M. Kafesaki, E. N. Economou and C. M. Soukoulis, Electric coupling to the magnetic resonance of split ring resonators, *Appl. Phys. Lett.* **84**, 2943–2945, 2004.
204. A. Grbic and G. Eleftheriades, Periodic analysis of a 2D negative refractive index transmission line structure, *IEEE Trans. Antennas Propag.* **51**, 2604–2611, 2003.
205. C. Sabah, F. Urbani and S. Uckun, Bloch impedance analysis for a left handed transmission line, *J. Electr. Eng.* **63**, 310–315, 2012.
206. J. Valentine, S. Zhang, T. Zentgraf, E. Ulin-Avila, D. A. Genov, G. Bartal and X. Zhang, Three-dimensional optical metamaterial with a negative refractive index, *Nature* **455**, 376–379, 2008.
207. N. Liu, H. Guo, L. Fu, S. Kaiser, H. Schweizer, and H. Giessen, Three-dimensional photonic metamaterials at optical frequencies, *Nat. Mater.* **7**, 31–35, 2008.
208. N. Liu, S. Mukherjee, K. Bao, L. V. Brown, J. Dorfmüller, P. Nordlander, and N. J. Halas, Magnetic plasmon formation and propagation in artificial aromatic molecules, *Nano Lett.* **12**, 364–369, 2012.
209. M. Albooyeh, S. Tretyakov, and C. Simovski, Electromagnetic characterization of bianisotropic metasurfaces on refractive substrates: General theoretical framework, *Annalen Der Physik* **528**, 1–17, 2016.
210. C. L. Holloway, E. F. Kuester, J. A. Gordon, J. O'Hara, J. Booth, and D. R. Smith, An overview of the theory and applications of metasurfaces: The two-dimensional equivalents of metamaterials, *IEEE Antennas Propag. Mag.* **54**, 10–35, 2012.
211. C. Tserkezis, Effective parameters for periodic photonic structures of resonant elements, *J. Phys.: Condens. Matter* **21**, 155404, 2009.

212. C. Tserkezis and N. Stefanou, Retrieving local effective constitutive parameters for anisotropic photonic crystals, *Phys. Rev. B* **81**, 115112, 2010.
213. X.-X. Liu and A. Alu, Homogenization of quasi-isotropic metamaterials composed by dense arrays of magnetodielectric spheres, *Metamaterials* **5**, 56–63, 2011.
214. A. Lavrinenko, A. Andryieuski, S. Ha, A. Sukhorukov, and Yu. Kivshar, Bloch-mode analysis for retrieving effective parameters of metamaterials, *Phys. Rev. B* **86**, 035127, 2012.
215. A. Lavrinenko, A. Andryieuski, S. Ha, A. Sukhorukov, and Yu. Kivshar, Bloch-mode analysis for effective parameters restoration, *AIP Conf. Proc.* **1475**, 140–142, 2012.

Index

ABCs, *see* additional boundary conditions
additional boundary conditions (ABCs) 9, 32, 39, 44, 95, 170, 212
anisotropic chiral media 77–80
anisotropy 63, 65, 67, 68, 77, 95, 101, 192, 232
antiresonance 30, 31, 34, 131, 238, 239, 246, 247, 253, 259, 265, 266, 272
antiresonant behavior 259, 260, 264, 265
approximation, first-order 119, 122
artificial magnetism 5–11, 30, 31, 42, 43, 79, 80, 89, 90, 102–104, 124, 128–130, 144–147, 160–162, 169, 170, 199–202, 222–224, 252–254, 269, 270

b-parameter 131, 133, 138, 140, 146–148, 154, 160, 163, 170, 186
bianisotropic media 42, 73, 75, 77–79, 81, 83–85, 103, 128, 136, 146, 155
bianisotropic responses 128, 145
bianisotropy
  artificial magnetism/resonant 16
  fictitious 264–267, 272

biisotropic media 73, 74
Bloch impedance 239–247, 267
Bloch lattices 241–243, 245, 247, 250, 253, 256, 257
boundaries
  effective 52, 67, 118, 119, 165, 170, 187, 188, 190
  effective-medium 188, 189
boundary conditions 9, 39, 170, 173, 174, 176, 178, 180, 182, 184, 186–197
Bragg phenomenon 12, 13, 25, 257, 258
Bragg regime 257, 258
Bragg stopband 13, 29, 227, 257, 259, 260, 267
Bruggeman model 86, 87, 90, 91
bulk homogenization model 17, 187
bulk medium 15, 16, 190, 196

canonical CPs 76, 78, 80, 82, 85, 113, 130, 153, 159
Cartesian multipoles 17
chiral media 4, 75–78, 80, 129
  natural 4, 5, 42, 80
  right-handed 78
chiral particles (CPs) 39, 76, 78, 84, 85, 89, 203, 263
chirality 2–5, 42, 74, 78–80, 82, 129, 138, 139, 264
chirality of natural media 4, 8, 42, 161

Index

chirality parameter 4, 74, 123
circular polarizations 76, 77
  clockwise 75, 76
Clausius–Mossotti–Lorenz–
  Lorentz formulas 46, 47, 49,
  51, 52, 56–61, 68, 70, 71, 148,
  169–171, 173, 175, 177, 179,
  181, 183–185, 187, 188
colored quartz 4, 5, 80
composite slab 18, 187, 188, 190,
  193, 196, 238, 249, 250, 252,
  253, 255
composites 2, 5, 6, 15, 19, 31, 41,
  42, 52, 70, 80, 85, 86, 88, 113,
  117, 154, 170
  molecular 2
continuity, effective 5, 9, 14, 16,
  18, 21, 26, 160, 170, 180, 185,
  247, 256, 262, 269
continuous media 6, 11–16, 21,
  22, 26–30, 32, 33, 35, 44, 45,
  50, 56–58, 147, 200, 201,
  208–210, 237, 239, 257
  electrodynamics of 14, 57, 269
  theory of 135, 239
coupling, magnetoelectric 34, 73
CPs, *see* chiral particles
crystal 44, 54–56, 62, 64, 65, 216,
  240
  natural 55, 58, 61, 63, 65, 137,
  193
crystal planes 12, 13, 53–56, 61,
  62, 65, 202, 206, 207,
  209–211, 213–216, 218, 240,
  245, 250, 257, 263, 271
  adjacent 12, 24, 25, 216, 230,
  240, 242
cubic lattice of spherical
  particles 67, 188

DC, *see* dichroism
dichroism (DC) 3, 39, 77–79, 82,
  83

dielectric medium 49, 133
dielectric spheres 89, 90, 153,
  160–167, 181, 187, 193, 195,
  211
  artificial magnetism of 160
  non-resonant array of 160, 161,
  163, 165, 167
  two-phase lattice of 223, 231,
  232
dimers 10, 31, 79, 80, 122, 132,
  133, 147, 229–231, 233, 247
dipole centers 100
dipole moments 24, 63, 94, 144,
  206
dipole polarizations 25, 53, 102,
  110, 155
dipole sum 48, 51
  planar 55
dipoles 47–49, 52–54, 57, 58, 61,
  94, 95, 141, 143–145, 148,
  151, 156, 162, 163, 165, 179,
  182, 271
dispersion equation 213,
  215–217, 219–221, 264
Drude layers 192, 194, 195, 271
Drude model 65, 66
Drude transition layers 66, 67,
  188, 190, 193, 196, 245, 270,
  271
dynamic averaging 199, 201, 203,
  205, 207, 209, 211
dynamic homogenization model
  31, 244, 247, 260

effective material parameters
  (EMPs) 11, 12, 14, 15, 17–19,
  23, 29–36, 43, 44, 73,
  104–106, 208, 209, 228, 229,
  231, 232, 237–239, 254–257,
  259–265, 271, 272
effective-medium model 25, 32,
  161, 166, 167, 169, 190
effective-medium parameters 270

effective-medium slab 30, 33, 186, 190, 236, 242
  layered 250
effective medium wave number 167, 173
eigenwaves 9, 24, 25, 75, 76, 94, 99, 137, 139, 202, 210, 211, 219, 242
electric dipoles 16, 28, 35, 39, 71, 107, 131
  z-oriented 81
electric field
  internal 89, 204
  local non-uniform 146, 151
  time-varying 59, 79, 82
  varying 10, 21
electric multipoles 10, 106, 107
electric quadrupole 83, 89, 104, 106, 107, 114, 116, 122, 125, 131, 132, 146, 149, 152, 158, 163, 180
electromagnetic field 10, 11, 20, 22, 75, 82, 93–95, 103, 135, 201, 236, 243
electromagnetic response 5, 6, 9, 24, 44, 83, 105, 137, 257
electromagnetic waves 2, 137, 147
ellipsoids 13, 14, 67, 68, 89
EMPs, *see* effective material parameters
  antiresonant 31, 231, 237, 244, 246, 247, 257, 259
  bulk 210, 220, 228, 234, 245, 250, 253, 256
  definitions of 154, 208
  dynamic 259, 263
  local 31, 34, 69, 95, 103, 137, 184, 213, 257, 259, 263, 271
  m-dipole scatterers 128
  retrieved 18, 30, 31, 36, 235, 237, 238, 246, 247, 256
  scalar 73, 101, 211
EMPs, Lorentzian 35, 229
EMPs retrieval 12, 36

Ewald homogenization model 58, 70
extinction principle 17, 35, 53, 57–63, 66, 170, 171, 187, 188, 190
extinction theorem 35, 56, 58, 59, 65, 66

ferrites 3, 10, 41, 75, 76
fields
  dipole 48, 54, 55
  macroscopic 58, 61, 95, 102, 112
  time-harmonic 73, 112
  time-varying 21, 183
frequency dependencies 236, 238, 246, 251
frequency dispersion 33, 41, 59, 115, 159, 225, 237, 238, 256, 264
Fresnel formulas 16, 56, 64, 66
Fresnel–Airy formulas 69, 235, 236
full-wave simulations 85, 91, 168, 195, 197, 213, 224, 228, 264

Garnett, Maxwell 68, 70, 71, 85, 86, 90, 91, 154, 165, 167, 169, 181, 195, 204, 225, 227, 228
generalized sheet transition conditions (GSTCs) 253–255
GSTCs, *see* generalized sheet transition conditions

high-order multipoles 42, 113, 139, 140, 159, 178, 180
homogenization 10–13, 15–17, 19, 25, 29, 49, 52, 53, 64–66, 68, 88, 95, 137, 139, 265, 269, 270

homogenization model 11, 12, 15, 17–20, 26, 27, 32 36, 42–44, 62–64, 66, 67, 70–72, 92–95, 100–103, 105, 106, 169, 170, 221, 256, 257
  complete 105
  first-principle 36, 37, 247, 257, 260, 263, 267
homogenization theory 15, 269
hyperboloid 13, 14

infrared SRRs, lattice of 236, 238, 245, 246, 252
interface
  continuous 58, 190
  effective-medium 36, 270
interface homogenization 15
interface problem 200
interference of partial waves 25, 257
isofrequencies 13, 14, 29, 222, 237
isotropic chiral media 73, 74, 77, 79, 80
isotropic media 7, 63, 73, 129, 195, 232

Kramers–Kronig relations 27, 29–34, 184, 185, 229, 256
  generalized 32
  violation of 30, 239
Kramers–Kronig relations for EMPs of media 33

lattice
  anisotropic 51, 67, 182, 185
  cubic 4, 48, 67, 68, 72, 91, 95, 148, 154, 177, 179, 196, 225, 247, 250, 251, 260
  dense 18, 57, 60–62
  finite-thickness 248, 251
  infinite 18, 36, 55, 205, 213, 221, 240, 244, 246, 255, 256, 271
  non-local EMPs of 32
  semi-infinite 52–56, 59, 200, 239, 240
lattice eigenmodes 32, 61, 62, 208, 222, 246
lattice geometry 95, 214, 215
lattice interfaces 18, 60
lattice nodes 93, 94, 105, 117, 186, 207
lattice of point dipoles 65, 190
local field approach 53, 60, 269
Lorentz, H. A. 15, 16, 48–52, 95, 96, 248, 250
Lorentz sphere 46–49, 51, 52, 58, 68, 71

m-dipole 71, 72, 81, 82, 104, 116, 129–133, 142, 144–146, 160, 161, 163, 165–169, 200, 201, 208, 209, 211–215, 217–221, 223–225
m-dipole grid 219
m-dipole moments 71, 113
m-dipole polarizabilities 165
m-dipole resonances 83, 152, 200
m-dipole responses 90, 133
macroscopic electromagnetic field 95, 97
macroscopic multipole densities 99, 109
macroscopic parameters 34, 66, 103, 157, 158, 184, 185, 245
macroscopic polarization current 97, 99, 101, 103, 105, 107, 109, 111
magnetic field
  tangential 217, 218
  time-varying 7, 82
magnetic multipoles 106, 107

magnetic quadrupoles 89, 106, 108, 114, 116, 121, 125, 128, 131, 158
magnetic resonance band 30, 90, 91
magnetic resonances 18, 134, 235, 237, 259, 271
magnetism, natural 6, 9, 20, 21, 75, 112, 135
magnetization 7, 10, 14, 74, 79, 102, 146
magneto-dielectric composites 68, 69, 71
magneto-dielectric media 79, 130, 133, 165, 237
  arbitrary anisotropic 209
  natural 10, 235
magnetoelectric coupling (MEC) 39, 73, 75, 80, 144
matched absorber 7, 8
material equations (MEs) 11, 34, 43, 73, 93, 94, 97, 98, 105, 106, 122–126, 128–130, 132, 133, 146, 147, 158, 186–188, 206–208, 270
Maxwell Garnett homogenization model 84, 91, 152
Maxwell Garnett model 68, 69, 71, 84
  generalized 127, 145, 154
Maxwell–Garnett model 158, 159
Maxwell's boundary conditions 6, 36, 199, 200, 235, 239
Maxwell's equations 51, 97, 112, 115, 122, 187, 205
MEC, *see* magnetoelectric coupling
medium p-dipole moments 108
metal bianisotropic particles, magnetoelectric coupling in 80
metal glasses 68–70, 85, 204
metamaterial frequency region 186, 187, 189, 191, 193, 195, 197

metamaterials (MMs) 6, 10, 15, 19, 30, 31, 91, 92, 199, 200, 212, 233–235, 237, 239, 247, 258, 267, 271, 272
metamaterials, homogenization of 199, 200, 202, 204, 206, 208, 210, 212, 214, 216, 218, 220, 222, 224, 226, 228
metasurface 197, 253–256
  effective 254, 255
microscopic field 15, 60, 100, 174, 199, 200
microscopic multipole densities 100, 106, 110
microscopic multipole polarizabilities 101
microscopic multipoles 101, 109, 119
microscopic response 29, 35, 42, 45, 106, 137, 159
Mie resonances 89, 90, 92, 169, 203, 226
MMs, *see* metamaterials
multipole 52, 94, 100–106, 108, 111, 116, 122, 134, 141, 151, 177, 187, 200, 205, 270, 271
  spherical 162
  static 153
multipole expansion 99, 101, 103, 105, 107, 109, 111
multipole hierarchy 42, 52, 127, 151–153, 200, 270, 271
multipole moments 83, 100, 101, 106–108, 111, 117, 126
multipole polarizabilities 17, 18, 83, 159, 163

nanoparticles 69, 70, 131, 204
near-field interaction 216, 218, 220, 230, 240, 242
non-locality 140, 141, 143, 145, 147, 149, 151, 153, 155, 157, 159

NRW algorithm 235, 236
NRW retrieval procedure 242, 244, 247

OA, *see* optical activity
OA parameter 76, 77
Omega-dimers 231–233, 258
Omega-media 78, 79
Omega-particle 39, 78, 82, 230, 233
optical activity (OA) 3, 39, 75, 76, 78, 79, 83
optical losses 8, 77, 91, 233, 256, 258–260

p-dipole 71, 72, 81–84, 89, 90, 100, 104, 118, 119, 132, 133, 144, 145, 152, 160, 161, 163, 165, 178, 211–216, 218
p-dipole grids 213, 216, 219
p-dipole lattice 217
  non-resonant 58
p-dipole Mie resonances 160, 265
p-dipole polarizability 49, 182, 183
p-dipole response 71, 211
particles
  bianisotropic 81, 83, 85
  constitutive 2, 5, 26, 67, 68, 82, 85, 100–102, 130, 146, 148, 151, 170, 171, 183, 211, 212, 269, 270
  dielectric 89, 160, 203, 204
  dipole 15, 49, 184
  metal 88, 143
  non-bianisotropic 266
  non-spherical 67
  p-dipole 105
  resonant 115, 171, 199, 203

permeability
  effective 7, 28, 34, 44, 90, 127, 128, 133, 134, 146, 161, 231, 245, 246
  magnetic 8, 10, 238
  resonant 134
permittivity
  bulk 191, 194, 255
  effective 7, 9, 31, 33, 34, 113, 115, 165, 166, 182–184, 187, 189, 190, 210, 211, 225, 226, 228, 244–246, 262
  negative 19, 160, 161
  non-local 9, 103, 135, 136, 212, 262
phenomenological approach 46, 67, 155
photonic crystals 2, 6, 12, 23–25, 29, 31, 32, 61, 222, 226, 229, 240, 257, 259, 264, 266
point dipoles 60, 61, 65, 66, 100, 102, 190, 206
polaritons 44, 60–62, 136, 199, 200, 243
  dense p-dipole lattice 60
polarizabilities 17, 18, 28, 37, 50, 54, 59, 61, 81, 87, 148, 157, 164, 185, 258, 269, 270
  individual 49, 85, 144
polarization 20–22, 45–48, 60, 61, 63–66, 77, 93, 94, 99, 102, 107, 109–112, 116, 117, 140, 141, 184, 185, 187–191, 207, 208
  bulk 28, 45, 47–49, 94
  counter-clockwise 3, 76, 80
  electric 22, 73, 81, 84, 90, 102, 145, 158, 174, 212, 229
  excessive 56
  macroscopic 14, 15, 21, 109, 207
  macroscopic medium 36
  magnetic 14, 73, 82, 155, 158, 229, 255, 262

medium unit cell 262
microscopic 11, 14, 15, 21, 55, 60, 100, 104, 107, 110, 174, 185, 188, 190, 206
p-dipole 114, 118, 148, 150, 185
polarization currents 11, 21, 112, 162, 163
polarization rotation 4, 76, 77, 161

quadrupoles 83, 113, 119, 122, 134, 142, 144–146, 149, 151, 156, 178, 179, 205
quasi-static homogenization 15, 19, 31, 42, 45, 69–71, 91, 92
  classical 35, 67
quasi-static model 15, 36, 86, 91, 166, 204, 208, 228, 232

reflectance 166–169, 193, 195, 196
reflection coefficient 56, 60, 66, 166, 235, 239
resonance 6, 7, 26, 31, 36, 80, 81, 83, 85, 89, 90, 166, 169, 171, 211, 212, 251, 257, 264
  Bragg 25, 256, 258, 260, 265
  magnetic Mie 89, 90, 92, 160, 161, 163, 165, 166, 204, 211, 223, 224, 259, 260
  plasmon 69, 70
  scattering 89
resonance band 34, 77, 80, 85, 86, 91, 133, 134, 145, 148, 161, 224, 227, 237, 256–260, 266, 271
resonance frequency 81, 82, 180, 204, 258, 259
resonance wavelength 82
resonant dispersion 225
resonant lattices 36, 204, 205

resonant losses 81, 226, 233, 237, 238, 247, 257, 258, 265
resonant MEC parameter 146, 264
resonant p-m-lattices 213, 215, 217, 219, 221, 223, 225, 227, 229, 231, 233, 235, 237, 239, 241
resonant stopband 229, 247, 257–260, 262, 264–266
retardation 3, 4, 10, 44, 55, 87, 137, 141, 147, 152, 173, 175, 177–179, 181, 183, 185

SD, *see* spatial dispersion
semiconductors 43, 44, 137, 161
  photovoltaic 43, 44
spatial dispersion (SD) 2, 5, 6, 8–10, 20, 23–29, 39, 41, 43, 114, 115, 124, 128–130, 134–138, 140, 259, 260, 262
spatial harmonics 134, 135, 205, 222
spheres
  core-shell 259
  electric 224, 225
  high-permittivity 91, 92, 113
spherical functions 162
split-ring resonator (SRRs) 7, 8, 10, 39, 72, 87, 116, 117, 130, 134, 146, 203, 229, 253
SRRs, *see* split-ring resonator
susceptibilities, multipole 109, 125, 126, 139, 140
susceptibility 45, 104, 125, 127, 131, 135, 157, 158
  m-dipole 121, 136, 158
  magnetic 123, 127, 159
  microscopic 28
  p-dipole 114, 115, 120, 121, 124, 136
  quadrupole 115, 119, 120, 124, 136, 158

tensors 37, 38, 59, 67, 68, 73, 75, 78–80, 101, 102, 107, 116, 117, 121, 123, 125, 127, 134, 135, 139, 140
  off-diagonal 79
transition layers 36, 64–66, 189, 191–194, 200, 201, 222, 248–253, 270, 271
transmission coefficients 16, 30, 56, 57, 64, 69, 193, 246, 249

wave impedance 79, 82, 208–210, 220, 235, 239, 240, 243–246, 250–252, 263, 266, 267

wave vector 9, 11, 14, 20, 23–26, 53, 55, 60, 132, 134, 136, 202, 205, 206, 209, 261, 262
weak spatial dispersion (WSD) 5–10, 18–21, 27, 35, 36, 39, 41–44, 102, 109, 110, 112, 113, 121–125, 127–129, 135–137, 160, 161, 165–169, 269–272
weak spatial dispersion, general introduction to 1, 2, 4, 6, 8, 10, 12, 14, 16, 18, 20, 22, 24, 26, 28, 30
WSD, *see* weak spatial dispersion
  theory of 9, 42, 44, 75, 106, 127, 129, 148, 151, 156, 160, 180, 187